THE TECHNICAL IMAGINATION

The Technical Imagination

ARGENTINE CULTURE'S MODERN DREAMS

Beatriz Sarlo

Translated by Xavier Callahan

STANFORD UNIVERSITY PRESS

Stanford, California

2008

Stanford University Press
Stanford, California

Library of Congress Cataloging-in-Publication Data

Sarlo, Beatriz
[Imaginación técnica English]
The technical imagination : Argentine culture's modern dreams / Beatriz Sarlo ; translated by Xavier Callahan.
p. cm.
Translation of La imaginación técnica: sueños modernos de la cultura argentina.
Includes bibliographical references.
ISBN 0-8047-8047-3542-1 (cloth : alk. paper)
1. Technological innovations—Argentina. 2. Inventions—Argentina. 3. Social perception—Argentina. I. Title.
T173.8.S267 2008
609.82—dc22 2007017226

Typeset at Stanford University Press in 10.5/12.5 Bembo

The Technical Imagination: Argentine Culture's Modern Dreams was originally published in Spanish in 1992 under the title *La imaginación técnica: Sueños modernos de la cultura argentina* ©1992, Ediciones Nueva Visión, Buenos Aires.

To José Aricó

Contents

Acknowledgments

I wrote this book at the University of Cambridge during the period when I held the Simón Bolívar Chair, peaceful months offered to me through the hospitality of the Centre for Latin American Studies (where I enjoyed the friendship of David Lehmann and David Brading) and King's College (whose faculty and staff kindly gave the most self-evident answers to my questions about age-old university traditions and the elements of a rigorous education). King's College provided a setting of near perfection for my stay. At the Centre for Latin American Studies—where Ana Gray took the trouble to make sure that nothing would trouble me, and where letters arrived almost daily from John King of the University of Warwick to further my education in the new customs—I worked on sources, skillfully compiled by Sylvia Saitta, which I had brought with me from Buenos Aires. I also worked on two brief preliminary drafts. One of these was read by Leandro Gutiérrez, by Luis Alberto Romero and Hilda Sabato of the Programa de Estudios de Historia Económica y Social Americana (PEHESA) and the Centro de Investigación Sobre el Estado y la Administración (CISEA), and by Adrián Gorelik, Graciela Silvestri, Anahi Ballent, and Oscar Terán. The other draft, which has since appeared in the review *Punto de Vista,* contained ideas that were the source of the videotape *Buenos*

Aires II, written with Adrián Gorelik and Graciela Silvestri, directed by Rafael Filippelli, and produced for a conference on utopia and the city held at the Club de Cultura Socialista. Carlos Altamirano read and suggested changes to the final draft of this book. All the people named here know what I owe them, particularly María Teresa Gramuglio, who granted permission to reproduce the painting by Juan Pablo Renzi that appeared on the cover of the Argentine edition; about Juan Pablo's death, in May 1992, there is nothing I can say.

Some months earlier, in August 1991, José Aricó had died. I would like to believe that he would have accepted, as amends for a relationship in which I was brusque and combative, my dedication of this book to him.

Translator's Note

In the Spanish of South America, the term *sector popular* (or *sectores populares*), like the Spanish adjective *popular* itself, presents certain difficulties for the translator who hopes to bring this term into idiomatic North American English. There is no question, of course, that *sector*/"sector" and *popular*/"popular" are direct cognates: Spanish *sector* is basically the equivalent of English "sector," and Spanish *popular* often conveys the sense of a thing's being broadly accepted and well liked, in the same way that English "popular," as in the terms "popular culture" and "popular opinion," often describes grassroots phenomena. Despite this obvious kinship, however, it takes only proximity to the unambiguous *sector*/"sector" to point up the separate roads that *popular* and "popular" have traveled, and the weight of the baggage that each has acquired.

Thus when the term "popular sector" is encountered in English, often in documents compiled or translated by nongovernmental organizations reporting on sociopolitical conditions in Latin America, it is generally an inverted, forced, unsatisfactory rendering of *sector popular*. The authors of such documents sometimes go so far as to place the literal translation "popular sector" between quotation marks and may extend the same treatment to other instances of "popular" in the English context—tacit acknowledgment that some trace of a foreign accent persists in both "sector" and "popular," especially when they migrate north as a pair. Up here, we have no trouble understanding what is meant when conversation turns to the public or private sectors, but the term "popular sector" seems to elude us. If it has a home among us at all, it tends to be in financial jargon ("popular sector stocks") or in a few other specialist lexica within our larger idiom. But it lacks the specific sense of *sector popular* as Beatriz Sarlo uses the term in depicting the Buenos Aires of the 1920s and 1930s, its vital plebeian neighborhood culture and the big dreams of its backyard do-it-yourselfers and crackpot amateur inventors, its lumpen cranks and autodidacts, as contrasted with that city's sober cadres of formally trained engineers, its haute bourgeoisie, and its lettered political and social elites.

To say that "popular sector" lacks the specific sense of *sector popular* is perhaps also to say that North American English, at least in the United States, specifically lacks—that is, excludes—a term equivalent to the Spanish one. This possibility invites further speculation, given our culture's long preoccupation with an ideology of classlessness, of universal access to so-called upward social mobility, including that ideology's need to minimize and thus reinforce what often turn out to be impassable social and economic barriers. (Not for nothing do our mass media routinely use the label "middle class" as a euphemism, a tool of erasure, in describing groups that could more honestly be referred to as the lower middle and working classes, the urban or working poor, the petite bourgeoisie and the proletariat—common or ordinary people, so called.)

Beatriz Sarlo's text is the story of how Argentine culture—literally, metaphorically, ironically—dreamed up, and dreamed itself into, an era of technological modernity. But it is also an extended dispatch from the class wars, from the front of idiosyncratic schemes, all quixotic, some explosive in every sense of the term, for self-propulsion above and beyond the battlefield. Sarlo confronts the battle itself unsentimentally, without artifice or bad faith. Therefore, where she uses *sector popular*, or where she allows *popular* to resonate as it can when readers have not been inoculated against questions of social class, the translator has tried to honor the spirit of her text by giving things their proper names in U.S. English, using "popular" only when it sounds natural, idiomatic, and intellectually honest—when it is, in a word, accurate.

Another, less intractable set of small complications had to be resolved in connection with the adjective *técnico/técnica* and with the noun *técnica*, particularly in connection with the issue of preserving the author's tone. Therefore, *técnico/técnica* has been translated as both "technical" and "technological," according to the specific context, and the noun *técnica* has been translated in several ways, again according to context: as "technology" or "technique" when those nouns serve both meaning and voice, and otherwise as "the technical" and "the technological" or, occasionally, as "engineering"—the very field to which so many puttering dreamers in Buenos Aires once aspired.

Warm thanks to Norris Pope of Stanford University Press for his perseverance and powers of persuasion; to Helen Tartar and Nathan MacBrien, formerly of the Press, for their early involvement in the project; to John Feneron and Rob Ehle for their beautiful work on the book's design and production; to the gifted Louise Herndon for editorial suggestions that improved the manuscript; and to Laura Pilnick for her spousal patience and profound generosity.

THE TECHNICAL IMAGINATION

Introduction

Every book comes from somewhere; this one takes up a loose and, no doubt, all but invisible thread from an earlier study.[1] There, in connection with Roberto Arlt, I spoke of poor people's knowledge, a motley blend of discourses on chemistry and engineering, metallurgy and electricity, exotic geographies, and visions forecasting the metropolis of the future. I made an attempt to learn more about this kind of knowledge because the preoccupation with technology that marks Arlt's fiction and other writings had also come from somewhere—perhaps from new symbolic configurations linked to the realms of popular culture, configurations that, once having entered the past, are difficult to grasp. In other words, did Arlt's passion for technology explain Arlt alone—a single case, fresh and unique—or did it reveal a trail of other passions? Did Arlt permit the divination of a "not-Arlt"? And, inversely, would this "not-Arlt" lead me back to Arlt himself?

I had to make a detour. A number of stories by Leopoldo Lugones, published in the last years of the nineteenth century and the first years of the twentieth, although they are also "science" fiction, are nevertheless different from some narratives by Horacio Quiroga that appear to have been written a short time later, under Lugones's influence. And yet what separates Lugones and Quiroga is not the years between them (which make the variance easy to see, since time's passing always creates the illusion of change) but the type of relationship that each of these writers established with readers (Lugones thought himself superior to his own) and with what was not "naturally" literature. The masterly, oracular style that Lugones used in *Las fuerzas extrañas*[2] did not allow the "science" in his stories to be anything but the most superficial kind of plot element; Lugones's poetics, like his life, remained unaffected.

I thought about Quiroga's life, marked by a technical bent that announced itself in the grace and audacity of a new kind of dandyism, a form novel enough to reinvent the dandy as a mechanically gifted industrial pioneer. But if Quiroga distinguished himself from Lugones by incorporating technology into his life, both aesthetically and pragmatically, perhaps he was a harbinger not so much of Arlt as of the culture from which Arlt came to literature.

It was worth the trouble, then, to ask what had been taking place within that culture. The fact that my inquiry happens to have begun with literature probably made it possible for me to approach not only the question (always an open one) of how a society is described in the texts of its writers but also the question (in the now classic phrase) of what that society has made of its intellectuals, and of what its intellectuals have made of what their society has made of them. Thus I came back around to poor people's knowledge—what it was, where it had been produced and how it had been disseminated, the experiences on which it had been based, the sensibility and skills it had made possible, the hopes it had aroused, and the gaps it had filled. For the common people, but not for them alone, poor people's knowledge was also modern knowledge. What needed to be determined was how poor people's knowledge had come to be fodder for an imagination that was not exclusively literary.

As a figure, Arlt, like Quiroga, allows one to imagine a network of relationships between the middle and working classes, between the readers of mass-circulation newspapers and the intellectuals of elite journalism, between social commentators and ward heelers. One story that was making the rounds in 1927 or 1928, the period when the newspaper *Crítica* was launched, came up again and again as emblematic—as almost iconically meaningful—in the memories of those who had passed it on:

> Literary themes entered uncharted territory when Roberto Arlt turned to the topics that got him excited: "Before long, radiovision, like radiotelephony, will be taking over our homes. A recent experiment shows that it's possible to manufacture affordable equipment and make it available to every social class, even the poorest. And here's another sensational news flash: Anton Raab, the aviator, has signed on with Opel to be shot into space, in a rocket now under construction, and go to the moon. This bold astronaut has made a prediction: 'If I do go to the moon, I'll die,' since there's no atmosphere."[3]

This blend of probability and imagination, of possible feats and invention, of future technological developments and hypotheses having less to do with science than with science fiction, was typical not only of literary discourse but also of the mass-circulation newspapers. It was an admixture that involved the impact (an important one, I believe) of technology as an instrument of economic modernization, and as a force for urban change, but also as a matrix from which ideal configurations of images had sprung forth, and from which there had occurred an unleashing of processes involving imaginary constructions no less than the acquisition of demonstrated knowledge. And it was more: one condition of this new knowledge's dissemination may have been the strong presence of a fanciful element, one that defined and hypothesized the future from the standpoint not only of science but also of scientific myth.

Reconstruction of the Imagined

How and why should attention be focused here? To this question, the chapters that follow are meant as a reply in six episodes.

First, however, this question presupposed another, which touched on elements of the imagination that were not completely developed in the Argentina of the twentieth century's second and third decades.[4] The antecedent question concerned particular aspects, neither political nor ideological-political, of working-class neighborhood culture—that is, the urban culture of Buenos Aires, created by Creoles and immigrants, and influenced by the processes of urban transformation, literacy education, and economic modernization. Needless to say, it concerned features that were novel.

The very notion of studying symbolic dimensions presented a series of problems. What image-repertoire could be considered in this area? Where should its discursive, graphic, or textual representations be sought? How should one judge its dynamics, its incidence, and its continuity in defining attitudes, hopes and expectations, identities,[5] trajectories, and myths?[6] This problem spilled over into questions about the degree of reliability that could be attained through the reading of assorted records from the past, and also about the reliability of the records themselves, oblique sources that know more than they tell but never say in so many words whatever it is one is looking for.

Likewise, a whole array of hypotheses was opened up by the question of how feasible it might be to generalize from those features that had made their way into the written record. It is impossible for these novel elements, which were personally experienced forms of modernization, and which for that very reason had the character of a cultural avant-garde, to have existed with the same productive intensity in every material and symbolic manifestation of the popular imagination. These were emerging features that did not yet make up a continuum of meanings; rather, they were conglomerates, with high mythic content, that made it possible to integrate the technological changes—in transportation, in communication—taking place in urban settings. But they were isolated instances, frequently unsystematic groups of motifs that retained their fragmentary character. What I sought was not, and perhaps never had been, a global tone but rather a new accent, already being heard at the beginning of the 1920s or even before, and becoming firmly established toward the end of that decade.

Obviously, in this as in every other case involving cultural innovation, the receivers and transmitters were a core group of the avant-garde: journalists or writers at the fringes (and sometimes at the center) of the intellectual world, young people taking classes or laboring at jobs, and men whose work brought them into contact with machinery, in addition to people at the fringes of high culture who saw in technology possibilities for social renown or social climbing not offered by other avenues.

At any rate, the lay use of technology was bringing about the reorganization of a hierarchy of knowledge; and literate artistic culture, which had not incorporated technology as a central motif, could now be appraised by outsiders right along with the contents of mass-market books, newspapers, and magazines. Technology made up for the knowledge and skills that one might lack in other areas. It had a dual purpose: cultural modernization, on the one hand, and compensation for cultural differences, on the other. Engineering (or what passed for it), the scientific notions published in technical manuals, newspaper articles—all of these stood in for the science taught in universities, the knowledge of the lettered elite, not replacing them but conferring on the culture that personified technology and scientific notions the same respectability and esteem enjoyed by more traditional kinds of knowledge. Such sources and notions may be viewed as ersatz, but that would be too simple a perspective; rather, they occupied an intermediate zone: they bridged a gap, and in so doing they validated, on one side, the importance of what was missing and, on the other, the legitimacy of a nontraditional reorganization of culture. As technologies, they belonged to the more familiar area of the trades, but as *new* technologies, they opened up a vista of original achievement.

I also entertained the idea that the culture industry, in the form of print journalism, had played an important role in a city like Buenos Aires in the 1920s and 1930s, both in defining plebeian culture and in discovering its more or less precise elements so as to generalize from them and feed them back to their social vectors. I confirmed my hypothesis that cultural producers and organizers who belonged to the educated sector (although not to the center of the intellectual world), and in some cases to the communications elite, had a finger on the pulse of that less accessible and no doubt less transparent domain that we usually call the domain of the common people.

This was true not only of Roberto Arlt but also of those who made up the journalistic elite of the big modern newspapers, which were defining new textual forms and new methods of reporting and understanding the news. I believe that it is possible to uncover a network of relationships between cultural consumers and cultural producers in the city of Buenos Aires, a network supported by institutional as well as noninstitutional mechanisms. I believe that these relationships presupposed a relay system, originating with the journalistic elite and passing through writers for the mass-circulation graphic media and writers whose work was published in the tabloids, finally reaching the aspiring writers, playwrights, pundits, and editors who were often very close, culturally speaking, to the new audience in the working-class neighborhoods.[7] The mass-circulation newspapers of the period speak of readers' and journalists' interest in the technological domain, certainly the domain of modernity but also of upward social mobility and cultural change.

It is difficult, of course, to marshal all the proof that might confirm these ideas. Among several possible avenues, I chose that of the popular press, then undergoing a period of consolidation and encompassing dozens of magazines, newsletters, catalogs, handbooks, and manuals. *Crítica* had the widest circulation of its day, and an examination of that newspaper from 1913 (the year it was founded) to 1930 indicates a rhythm of rapid (or visionary) adaptation, and a broad scattering of themes. As for *El Mundo*, a morning paper also aimed at the middle and working classes, it had been using this kind of material to compete with *Crítica* since 1928, the date of its own founding, contemporaneous with the publication of Arlt's novels. The articles in *El Mundo*, heterogeneous in nature, were an informational mélange of highly uneven reliability. In short, it was "anything goes" in this jumble of science, vulgarization, invention, popularization, simple and simplistic explanations, weird news of the type that *Caras y Caretas* had been publishing from its beginnings in 1898, profiles of inventors, regularly published sections that came and went, images of near and distant futures and of the far side of the moon, aviation and interplanetary travel, television and telephony, geographical discoveries and explorations, intersections of parapsychology with quackery and superstition,[8] technology applied to everyday life, military technology, miracles, and miracle cures. I also traced the spread of these cultural motifs through evidence provided by advertisements, pulp catalogs, correspondence courses and the numerous pamphlets advertising them, lectures, patent registrations, inventors' societies, and groups of ham radio operators.

Clearly, this work's guiding concern begs the question—one that all of us today have a stake in answering—of which possibilities were open in early-twentieth-century Argentina, a country that certainly reversed its course quite radically in the 1940s. Perhaps all this interest in science and technology also involved a fascination with novelty and transformation, the impetus and foundation of the aesthetic vanguards. Thus the public universe spilled over into the universe of the intellectual movers and shakers, with a resulting continuity of values, once again highlighting the importance of communication networks for the journalistic-intellectual elite and its audience of common people—or, to put it more cogently, their importance for the communication of different cultural dimensions.

Gaining a Culture

Plebeians, especially those who were immigrants, carried out complex and more or less successful operations of assimilation into a "common" culture. If the educational establishment laid down the conditions and rules for the "Argentinization" of immigrants' children, and if these children and their parents envisioned an assimilation broadened and enlivened by Creole customs,[9] then it is also possible to understand the interest in technology, and the fad of technology, as alternative strategies for cultural integration, concerned equally with immigrants' assimilation into the dominant culture (defined by the state and by the intellectual and journalistic elites) and with the establishment of the immigrants' own social variants within that culture.[10]

The new culture of technology, which in the West was undergoing a period of expansion and spectacular unfolding, had an impact at the very point where cultural permeability was most in evidence, and where no one was sure that there actually was any past to be defended against the progressivist wave: among the common people who were recent arrivals to literate culture, hard-pressed to assimilate its rituals and practices and seeking cultural forms that, without becoming completely alienated, could transcend the ordinary limits of working, of producing and reproducing material life. The place occupied by sentimental serial novels, those relatively respectable dream machines, could also be occupied by a technical or, rather, technographical imagination, one that viewed problems, technical and nontechnical alike, from the starting point of a small kernel of novel elements—a popular "science" that basically presented itself less as knowledge than as know-how. And I also use the term "technographical" because this kind of imagination enables one to speak of other things that do not belong to the realm of technology; the term is useful in making comparisons, analogies, and metaphors. From another perspective, one known in the United States since the days of Ralph Waldo Emerson and Walt Whitman, technology's function was not limited to merely utilitarian applications; technology also offered aesthetic inspiration along with scenarios in which new systems of objects were unfolding. To the early-twentieth-century Spiritist reaction in Río de la Plata, the introduction of technology responded with a local patchwork of ideas and attitudes that in the northern hemisphere, from Europe to the Soviet Union, were ascribed to "Americanism."[11]

In the 1920s, this process was accompanied by technological changes whose cultural resonance was very strong: electrical long-distance communications media were introduced into Argentina at almost the same moment as in the advanced Western nations. The results included a virtual radiomania, great expectations for acquiring skills, the importing of gadgetry, and, in particular, an incentive for homegrown research and development. Radio was an obsession among the local technicians, but not just among them or the elegant ham operators who had founded the Radio Club. In August 1920, with the legendary broadcast of *Parsifal*, radio made an explosive entry onto the Buenos Aires scene.[12] Radio's impact had to do with the medium's possibilities, at once technological and cultural; indeed, it had to do not only with a spectacular innovation but also, strictly speaking, with a new cultural genre. As an innovation, radio was the fulfillment of fantasies that were not strictly technological: wireless long-distance communication; the harnessing of invisible waves, and their reception on primarily galena-based devices; the presence of disembodied voices and music, which evoked dematerialization and movement away from a culture rooted in unmediated vision and toward a culture sustained by mediation. In an era preoccupied with telepathic transmissions, hypnotism, and the reception of supernatural messages (as further shown by the newspapers' intense interest in these topics, and by the growth of the Spiritist schools), radio was a cultural revolution, given what it directly represented not only as a communications medium, and as the domain of the industrial mass-media culture that was to flourish in the 1930s, but also (and, I would say, more profoundly) as a miracle of technology: the material means of making possible what hitherto had been impossible. Toward the end of the 1920s, the sound of radio had been incorporated into urban space, and it had become a presence in the cultural landscape of Buenos Aires, contributing to changes in the perceptual system with a speed, penetration, and persistence never before enjoyed by any innovation in communications technology.

If radio was a miraculous invention, technicians and hobbyists were kept busy with other inventions of the wildest and most trivial sort. It was in the imaginations of journalists, businessmen, and skilled artisans that the very figure of the inventor as a modern type took root. Invention, an avenue promising the double denouement of riches and fame, was a cultural motif of the 1920s, whether found in the works of Roberto Arlt or traced through magazines, newspapers, and the Office of Trademarks and Patents. Understood as a stroke of fortune, invention offered a model of upward social mobility based on nontraditional skills. First for Horacio Quiroga, and then for Arlt, invention was a theme in the most definitive sense: it served as an organizing principle for fiction, it reconstructed a hierarchy of values, and it offered freedom from the constrictions of the genteel tradition of the entrenched elites in literature and the fine arts.

In this way, through the miracle of modern inventions (not to mention frauds and hoaxes), I managed to find another way back to Arlt. In these six episodes of Argentine culture, I have compared and contrasted literary and journalistic texts with whatever traces of practices and subjectivities can be inferred from another body of heterogeneous writings. To compare and contrast is not to explain everything; it is simply to say that literature crossed paths with real life in the most creative registers of a new cultural dimension.

This new dimension demands that we reconsider the profile of modernity in Argentina: its weaknesses (in the double sense of what was missing and what was desired), and its power in the coming together of a number of writers, dozens of journalists, and a public that found in modern technological fantasies the very stuff of fascination.

PART I

LETTERS

CHAPTER ONE

Horacio Quiroga and Technoscientific Theory

> Ricardo Ortiz, an Argentine, was born in the federal capital. His family, a very rich one, decided that he should study electrical engineering, for which he had demonstrated a strong inclination since childhood. He was brilliant at his studies in Buffalo. Upon returning to Buenos Aires, instead of pursuing his profession, he devoted himself to the study of electrical batteries, believing himself to be hot on the trail of a new, incredibly strong and stable element.
>
> —Horacio Quiroga, *El hombre artificial*

Biographical Data

Two friends, concerned less about supporting a thesis than about cranking out the "life" of a writer in which everything would have its place, even those traits that the mores of the time might have found less than acceptable, offer the readers of their book, written in the months after Horacio Quiroga's death, numerous bits of information about his technological passions.[1] No fewer than twenty times, in a volume of four hundred scant pages, they refer to the biographical subject's experiments, his workshops, and his technical failures and fancies. These mentions seem less likely to have been derived from research than to be simply unavoidable when the authors refer to the various houses occupied by Quiroga (dominated by a chemical laboratory, an electroplating works, a ceramic oven), or when they cite the equipment with which he set off for the province of Misiones, or the physical labor invested in the hinterland from which his second wife fled in boredom, or the ventures he attempted to launch there in order to free himself from the burden of poorly paid writing for newspapers and magazines, or the enthusiasms of his youth and early middle age, first for the sport of cycling and later for his motorcycle and then for a boat that he built himself and finally for a Model T Ford.

What the two biographers, José M. Delgado and Alberto J. Brignole, have constructed from this information is nothing more than a psychological profile showing Quiroga's tendency toward a "pleasure complex" encompassing physical activity and mechanical challenges. They fail to connect the dots between the facts that they provide, and so each fact takes on singular importance, since each one's presence is at once inevitable and justified only by a biography that cobbles together the themes of the Quirogan myth. But one of these themes—experimentation and technological trailblazing—is a nontheme, simply there, unworthy of comment.

If Quiroga, while still in Salto, Uruguay, and before the age of twenty, "exhibited any predilection apart from his disorderly zeal for reading, it was not for the liberal professions but rather for the trades." Chemistry was soon added to mechanics:

> His rooms in town and in the country were turned into laboratories outfitted with all manner of retorts, manometers, alembics, and flasks filled with the widest variety of alkalies and acids. He spent long hours holed up inside, repeating basic experiments of analysis and synthesis. But his imagination, incapable of resigning itself to this passive, routine role, often tempted him to actually conduct the totally unheard-of tests that it had proposed to him.[2]

Next, predictably, came photography, considered "more trade than art":

> The cycling magazines, the images of cycling champions, and the chemical library and arsenal were joined by collections of photographs, sets of enameled trays, fixatives and developers, cameras, and, in one corner, a camera obscura.[3]

When Quiroga arrived in Buenos Aires, and for several years afterward, he shared an apartment with Alberto Brignole. What his friend recalls of the room occupied by Quiroga is a scenographic amalgam of activities and implements, but especially of tools for Quiroga's newfound enthusiasm (to which Roberto Arlt, in particular, was also no stranger): galvanization and electrolysis, the chimeras, in modern form, of matter's conversion and reversion.[4] This enthusiasm was both technological and aesthetic, based as it was on the transformation of matter, and especially on matter's enrichment: through electrolysis, the basest metals would be turned into brilliant, lustrous materials closer to gold and bronze, and with the superficial nobility of the rich substance that had, in the process, been created and attached to them. At the same time, however, electrolysis was the modern form of alchemy; it appealed to the imaginations of those who would transform matter, and in those years it had artistic appeal along with obvious practical importance.

Quiroga, strictly speaking, became as actively familiar with technology as with the trades. He represented something new in the Platine cultural universe, something that did not just involve actual proximity to materials and tools; it also involved the bridge (built through the books and technical magazines he was reading[5]) to a skill set that was neither intellectually distinguished nor part of any broader local tradition among the lettered elite. Occupational know-how probably figured in almost all of Quiroga's escapades, from his first cotton business in Chaco Austral to the more or less exotic botanical gardens that he managed to plant, graft, and hybridize toward the end of his stay in San Ignacio, Misiones. In between were the "inventions" (his biographers call them "chimerical undertakings"), which are usefully viewed not just from the standpoint of a writer in the missionary jungle who was trying to survive outside the literary marketplace of Buenos Aires but also as strategies for using technology and technological skills to establish power over nature. Quiroga fantasized about a number of innovative enterprises, which he ascribes to one of his missionary characters: "The manufacture of cracked maize, always scarce in those parts; mosaics of tinplate and ferruginous sand; nougat made from peanuts and honey; extracted incense resin; glazed orange peels, samples of which had driven the maté harvesters mad with gluttony; trumpet-vine dye, precipitated by potash and essential oil of orange"; and the list grows longer, with disinfectants, superphosphates, dyes, latex extraction, the construction of desiccators and rails, and the like.[6] In 1914, when World War I made European carbon imports impossible, Quiroga embarked on the manufacture of charcoal, a process that outstripped his financial resources as well as his technical expertise and came to its ironic end during the final firing, in a huge, rather foolish conflagration inside the furnace.

These basic projects of agrarian engineering went beyond the notion of a mechanical hobby or the reappearance, through a kind of rural industrialism, of urban dandyism, although the latter aspect cannot be disregarded. Indeed, the fervor of the pioneering technician owes something to the gambler who finds himself in difficulties because of bets whose outcome is out of his hands, even though he thinks he has some understanding (in this case, technical understanding) of the game. In the proclivity for financially risky adventures that bourgeois prudence would consider foolish, there is also the ironic distance of the dandy that Quiroga was. Finally, there is the literary taste for intense experience of a modern Robinson Crusoe making his own way down the path of invention and applied technical imagination, a writerly figure, utterly unknown in the River Plate Basin, who nevertheless evoked characters from Jack London: naturalism and philosophical materialism, in pragmatic terms.[7]

Quiroga felt the pull of technology and applied innovation, and this attraction was certainly, if not exclusively, determined by a naturalistic poetics. Before he wrote "El conductor del rápido," he resolved to take a train trip in the company of the conductor,[8] a goal perfectly consistent with the aesthetic-moral imperative of naturalism; but the pleasure he took in technology's most commonplace manifestations is also what led him to ride the Ferretería Francesa of Buenos Aires with the dedication and pleasure of the new type of flaneur, and to scour bookstores for artisanal and trade manuals with the zeal of a Benjaminian collector.[9]

If Quiroga was reading Sherwood Anderson, as accounts of his contemporaries report, then a kind of adventurous vagabondage through both space and matter can be discerned in these passions: he was finding concrete information in places and materials with which literature did not associate, but especially information of a type that, if not new, was nearly inconceivable as an element of a writer's intellectual education. The ideal of the man who can change his social status is not exclusively a myth of upward mobility; it also includes a change of status by way of the information contained in an itinerary that is independent of libraries. The Americanist inflection of this composite of ideas was a stroke of originality among the writers of postmodernism: as in the United States, technology could drive not just an individual's life but also the model of a society whose members were equals regarding information of a practical kind, information whose newness was in itself a leveling force.

Quiroga's obsession with mastering all the trades was not merely a psychological trait; rather, it was a moral ideal of independent self-construction, conceived in terms of the future. His love of speed, born with the cycling club he had founded in Salto, and with his thwarted career as a racer, went on to find its most contemporary emblems in his motorcycle (Quiroga, a true dandy, or a man ahead of his time, had bought one in 1918) and in his Model T (purchased in 1925). As a plebeian mechanic, Quiroga performed an operation of virtually constant disassembly and reassembly on the Ford, itself a bricolage of pieces obtained on pilgrimages (these can only be imagined) to factories and workshops or spare-parts junkyards, peripheral wastelands that, according to Roberto Arlt, were also frequented by amateur inventors.[10] The futuristic love of speed brought machinery under the sway of a constant struggle with material limitations and practical skills; both these elements of motoring and motorcycling were present in Quiroga. But there were also marks of turn-of-the-century dandyism in this cultivation of technical prowess, which sometimes brought about the possibility of erotic scenarios for Quiroga the San Ignacio trailblazer as well as for Quiroga the lover traveling by motorcycle from Buenos Aires to Rosario.[11] And his final activity in Misiones—high-tech "scientific" horticulture in his garden of orchids, amaryllis, and poinsettias—bears witness to the taste for exotic tropical flowers (sketched by nature as if from an Aubrey Beardsley drawing) that runs through modernism, art nouveau, and the *stile Liberty*.[12]

Quiroga was not alone in traveling the path of modernity, technology, and dandyism (compare Marinetti, D'Annunzio, and—somewhat later, in his own way—Oliverio Girondo),[13] a path that leads almost ineluctably to the cult of the cinema. It is well known that Quiroga, from 1919 on, had been writing newspaper articles about films and had built several stories around themes and plots taken from the world of motion pictures. We know that the earliest films (short subjects and serials) did not excite universal interest among intellectuals and artists, and so Quiroga's enthusiasm once again establishes him as a pioneer, and in this case his position can be explained both by his fascination with technology and by his being intellectually attuned to something new, for it was precisely the emergence of a new kind of audience that inspired the first of Quiroga's stories with a filmic theme, "Miss Dorothy Phillips, mi esposa," published in 1919.[14]

A Technoromantic Ideal

Motion pictures were the turn of the century's technological wonder. With the strong impact of the Lumière brothers' *Feeding the Baby* and *Watering the Gardener* (the first slapstick film), technological wonder joined forces with fantastic narrative.[15] Amazement in the face of the marvelous—as Méliès,[16] who was working with both of these inseparable elements, quickly grasped—was the product of increasing and, for all practical purposes, perfectly limitless refinement in the mechanics of trick photography. Thus movies called to Quiroga from two domains, one evoking technical or formal possibilities, the other belonging to the register of the imagination, that linked two poles of aesthetic desire at the beginning of the twentieth century.[17] Motion pictures offered a new setting for the literature of the fantastic; where Quiroga's poetics was concerned, cinema allowed still unknown creative possibilities to be grounded in technological developments.

This is precisely what is articulated in stories like "El espectro," "El vampiro," and "El puritano," where motion pictures are a theme in the most literal sense (the characters are movie actors and actresses or are associated with them) and at the same time form the basis of what is being played out in the fiction. In their thematic function, movies restocked the sentimental image-repertoire and shaped eroticism in a radically new way: in these stories (and also in "Miss Dorothy Phillips, mi esposa"), passion and the stages of falling in love, which sentimental literature had made a powerful force in the literature consumed by a plebeian readership, are pushed toward a frenzy of love and death that evokes late-romantic and decadent ideals. But Quiroga did not rewrite these ideals in a literal sense. On the contrary; motion pictures, in their second, technical function—as a formal principle of narrative, a device at once narrative and technical—established a certain distance from the thematic. This distance was in its own way ironic: an exaltedly romantic narrative imagination hinging on the theme of "love beyond the grave" was made possible through the preeminent invention of the modern era.

These stories' cinematic settings undergird a narrative practice tinged with the classic themes of fantastic literature. How can the lover gain possession of the beloved's image? How does this image attain an embodiment that turns it into something more real or more powerful than life and death? These are the questions posed by Edgar Allan Poe's "The Oval Portrait," and Quiroga had already tested their narrative potential in his own "El retrato,"[18] but the cinematic setting allowed him to develop new possibilities in at least two ways.

First, Quiroga intensified the promise that motion pictures held for the fantastic, as a technology for producing and reproducing images. If a moment could be captured for all time and conjured up at will, if the two-dimensional static photographic image, though still flat, could be temporalized by motion, and if the pure present of the image could in fact be the harnessing of an infinitely actualizable past, then, at least in theory, one could not dismiss outright a technological development that might allow passage between the two-dimensionality of the image and the three-dimensionality of the world, between the frozen present of the image and a motion that might set it free from repetition and reinscribe it into the flow of time. Quiroga's stories presuppose Morel's invention; Morel's invention elaborates a proposition (as Quiroga's stories also began to do) concerning the supposed potential of motion pictures to create image-realities.[19] Ghosts and vampires are projections of an image that has been technically perfected to the point of crossing the boundary between the cinematic analogue and the primary referent (that is, what the camera has captured on film).

The hypothesis that it might be possible to pass from two-dimensionality and repetition to three-dimensionality and temporal flow proceeded from an analogy that rested on photography, which stood at an earlier technological phase. If three-dimensional reality could be captured on a flat surface, then it should have been possible (as motion pictures had demonstrated) to free that surface from its primary immobility and then from its prison of temporal repetition (and this is the technofictional presupposition of Quiroga's stories). The rays imprinting a negative were not the most valuable ones that a technical procedure might be in a position to control; it was possible to imagine other rays that could recover and liberate the "life" of the imprinted images. And motion pictures might be able not just to reduplicate a referent but also to create a reality that was comparatively independent of the original image. Such intellectual games, played with the laws of a brand-new technology, were at the root of hypotheses that, like the ones in science fiction, built up an *as if* on the basis of technical or scientific extrapolation. And here, precisely, is the foundation of Quiroga's stories: his narratives function *as if* it were technically possible for movies to bring about the viewer's (or the protagonist's) dream of blending motion pictures with life, of carrying the filmed scene's strong emotions forward onto the stage of reality.

The second way in which a cinematic setting allowed for development of new possibilities had to do with a more straightforward reflection of the emotional realm, as mediated by the technician. Here, strong feelings had the power to vanquish death; a cinematic image pulled from the screen could suck the blood of a living man; a dead husband's jealousy was able to modify images in the film within which he, as an actor, watched and was watched by his wife and her lover; a man's love for an actress might succeed in trapping her image, extracting it from the celluloid, and reconstructing it as a luminous body in motion against an actual scenic backdrop. In short, erotic desire manipulated the laws of technology and, in the end, fell victim to that manipulation.

In order for these stories to be written, these two aspects of cinema, the erotic and the technological, had to intersect. Quiroga captured and was captured by both: the erotic potential of the cinematic image mattered as much to him as did its potential for generating settings that were at once fictional and technological. The intersection of these two aspects is what makes the narratives what they are, fantasies that owe as much to technology as they do to Eros: the vampirism of rays of light captured on celluloid, or Faustian myth based on the latest technology. The fantastic, which involves the potential for setting an image free, is made possible by technology, which allows the image to be capured and indefinitely reproduced; the breaching of technological limits—or, to put it another way, the marshaling of technology by obsessives, lunatics, emotional cripples, or dunces—creates tragic outcomes that contain a twofold moral.

On the one hand, modernizing tensions elicit a technology without material or ethical limits; on the other, material forces avenge themselves on the apprentices who handle them. If cinematic technology encouraged the belief that anything was possible, the results of the actions unleashed by this idea show the sinister aspect of technological extrapolation, an aspect that Quiroga links to archaic phantasms of hysteria and vampirism and to archaic laws of guilt and vengeance. That is exactly how one genre of cinema worked in the first decades of the twentieth century, approaching various romantic or sentimental topics with the most up-to-date technological resources, and exploring a new kind of formal narration with materials taken from antiquated sources that literary narration had left behind.

Quiroga's attempts, in these stories, to splice late-romantic mythology with sophisticated technology took place in an environment where cinema had already had a powerful introduction to the wider public as a mode of aesthetic sensibility, and to some narrower groups as a technology-based hobby. In the latter connection, one can cite the earliest cinematic experiments conducted in Buenos Aires as well as the record compiled by such major newspapers as *Crítica* about motion pictures and about the proliferation of magazines specializing not just in the mythology of the star system but also in technological advances and tricks of the trade. In Buenos Aires, the year 1919 saw the appearance of *Imparcial Film*, a magazine devoted entirely to motion pictures; in 1920, *Cinema Chat* and *Hogar y cine* began publication, followed in 1922 by *Argos Film*, by *Los héroes del cine* the following year, and then, in 1924, by *Film Revista*. By the mid-1920s, weeky fiction magazines included sections on Hollywood, with two principal motifs: lives of the stars, and industry gossip. Local inventors patented several early improvements on the technique of capturing "moving landscapes," and they competed with renewals of foreign patents.[20] *Caras y Caretas* frequently advertised not just photographic equipment but also movie cameras and projectors for film enthusiasts. And in this collective drive toward the technological reproduction of images, which certainly had already found its audience, Quiroga himself eventually took part: not only did he have a photographic laboratory, in the first decade of the twentieth century he also accompanied Leopoldo Lugones as the latter's photographer on a trip to the Jesuit missions. The power of technological wonders had captivated only a few intellectuals, but it had already taken root in the imagination of Buenos Aires.

The Primitives of Technology

Motion pictures had pushed technological issues to the forefront of innovation, but Quiroga's involvement was not limited to these radically new areas. He was also fascinated by the primitives of technology, artisans who had a craftsman's manual dexterity but sought to apply it to machinery. Quiroga himself was one of them, and, like them, he prided himself on the perfection of his work with wood, ceramics, or metal, at a time when he was envisioning what would in fact always be an imperfect control over processes of production that lay beyond the limits of artisanal skill. There is something tragic about the struggle to achieve impressive outcomes in mechanical engineering from a starting point of sketchy information and precarious material circumstances; it is precisely this conflict that is laid out in stories like "Los destiladores de naranja" and "Los fabricantes de carbón," which appeared in 1926 and 1921, respectively.

The primitives of technology are putterers, for neither the materials nor the machine parts they use are adequate to the functions they should be serving in the invention of new processes (always defective) that mimic those normally used in industry. The primitives of technology construct stills in which "every serviceable fragment [has] been replaced by some substitute," in a functional transmutation that can only lead, through incomplete, imperfect approximation, to failure. Imitation by substitution, or the characteristically plebeian reinvention of the wheel, has the quality of a heroic exercise of patience because the final result always exhibits its humiliating difference from the idea or model. The practical objective of the invention (in the stories just cited, the manufacture of orange brandy or charcoal) is constantly postponed by intermediate steps—how to get an improvised boiler to work, how to connect the pipes of a still or fit rivets when there are no rivets to be had—that represent partial triumphs or partial failures. These intermediate steps become accomplishments in themselves and, in the end, obstacles that prove to be definitive because the amateur inventor never completely manages to overcome them.

The gadget manufactured by a primitive of technology is a deformed imitation that tinkering has made into a chaos of superfluous duplications and critical lacks. In conditions of uncertainty, technological imitation becomes a baroque outburst of augmentations, paste-ons, amendments, and wrong answers imposed by the material conditions in which a problem has been formulated:

> Major frosts had caused the oranges to ripen very early that year, and One Arm also needed to give some thought to the temperature in the storehouse so that the chilly nights, still very much with them in October, would not interfere with fermentation. Thus he was forced to insulate his shed with handful after hand-

> ful of sprouting potatoes, until it looked like a bristly, aggressive brush. He had to install a heating device, with a firebox made from a disinfectant drum, and bamboo tubing that coiled around the shed's straw walls like a fat yellow snake.[21]

And the charcoal manufacturers construct their boiler, the mechanical heart of the process, through another system of substitutions, one that intensifies their activity without offering any guarantees about the final result:

> With this—four pieces of sheet metal left over from the assembly of the shed—and the help of Rienzi, they could have a go at it.
>
> And so they did. These materials sufficed because in wood extraction, coal gas does not work through pressure. Using T-shaped iron braces for the mounting and L-shaped ones for the openings, they hung the rectangular boiler, 4.20 x 0.70 meters in size. It was tedious, tough work: in addition to the technical problems, they had to reckon with those caused by the lack of materials, or of this or that tool. For example, the initial coupling was a disaster: it was impossible to fasten those flimsy edges together, much less do it in midair. So they had to fit the boiler out with rivets, one per centimeter, which meant 1,680 of them just to fasten the lengthwise sheets. And because they had no rivets, they cut down 1,680 nails, and a few hundred more for the mounting.[22]

The charcoal furnace is equally provisional. The impossibility of regulating the temperature inside the boiler, the fragility of its walls, catastrophe in the person of a laborer who has no technical expertise whatsoever, whereas his bosses, the partners in charcoal manufacturing, think they know it all—these factors ultimately come together to create a total breakdown in the circuitry. This is, in effect, on-the-job training in that no-man's-land where mediocre education meets pioneering adventure; as Quiroga aptly describes the two men, "they dressed like laborers and talked like engineers, but they were not engineers, they were not laborers." As a result, they jot down only the roughest of notes while testing their system, not to mention the fact that winter temperatures at the mission—which have been, as it happens, unusually cold—almost always hinder calculations of any kind. The improvised nature of the construction (nails passing for rivets, leftover sheet metal for the boiler's walls, wire and clay not quite managing to become insulation), together with the two partners' obsessive assumption of the drudgery and physical discipline to which they submit themselves, constitute a system of all-encompassing incongruity. They are, ironically, professional practitioners of a dubious, inept mechanical engineering, one that penury is able to transform into a recursive loop of suffering from which the only possible escape is the strength of a naive optimism—the psychological cliché that defines One Arm in "Los destiladores de naranja."

One Arm is almost too perfect a match for the typology of the indigent amateur inventor. In his lack of financial capital and the inadequacy of his information, he recalls the dreamers Arlt found scavenging the junkyards and beaches where mechanical and industrial refuse piled up in the Buenos Aires of the 1920s and 1930s. With his indestructible faith in the potential of new applications of technology, One Arm is also reminiscent of the amateur inventors who used to drop in on the editorial offices of the Buenos Aires newspapers. But he is all these things in Misiones and is therefore even more cut off from essential technological resources than the plebeian amateur inventors of Buenos Aires. His strongest tangible links with technology are a soldering iron and, in a truly hilarious symbolic touch, two volumes of the *Encyclopédie*. In One Arm's world, the *Encyclopédie*, with its handsome plates illustrating techniques that have already undergone a century and a half of accelerated innovation, is a resource at once anachronistic and inaccessible, for it is completely alien to the poverty- and ignorance-bound environment in which One Arm is literally a mad scientist, oscillating between replication of a well-known procedure (distilling brandy from oranges) and a chimerical work of engineering that is baroque in its pointless complication ("using filtration to divert the water from the Horqueta wetlands to his house").

As tinkering artisans, both One Arm distilling his oranges and the two friends designing and building their charcoal furnace reveal the ambition and limitations of a technology that never quite rises to the challenge, simple though the challenge may be. The artisan's characteristic pride in a job well done shrinks in the face of a failure that calls to mind the limitations of any intervention based on nothing more than an artisan's expertise and materials. Quiroga found it interesting to relate the minutiae of these experiments, just as Arlt, in his novels, transcribes chemical formulas and designs for machinery, because the trailblazing of these amateurs and primitives carried symbolic weight in a world where new types of learning were changing the traditional organization of information and skills, at least among the common people.

Quiroga was aware of this change; a major part of his life, not to mention his leisure pastimes, was linked to it. He, too, was a naive builder, if you will, a technological pioneer (all the more so, in view of the gulf between technology and literate Platine culture that was characteristic of those years). Increasingly, however, it was inventors and replicators of inventions who were cited in the mass-circulation newspapers of the time, and the fact that they were considered newsworthy by publications attuned to developments of interest to the common people makes it possible to imagine the excitement aroused by even the most rudimentary approaches to manipulating materials and machinery, given the potential for learning and the promise of economic gain that could be realized in this way. Even when it appeared that a venture was not destined to be a success, the prospect of some future favorable outcome was not out of the question, in the eyes of these amateurs and the people around them; the five-year-old daughter of one of the charcoal manufacturers asks her father if he is going to "make a ton of money" with his new machine, and in the last line of the story she also consoles him for the recent failure: "Poor Daddy, your boiler burned up! . . . But don't be sad! . . . You're my favorite inventor in the whole wide world, and you're going to invent lots more things!"[23] Although the possibility of financial gain was not absent from these technology-based fantasies (the economic aspect of invention is fundamental in Arlt), they also had value in their own right.

Technological pioneering—a form of modern adventurism conceived, American-style, as a frontier battle in which the protagonist brandishes his practical expertise—furnishes Quiroga's narratives with a plot of conflict and suspense. In these two tales, about manufacturing charcoal and brandy, familiar story lines are combined with the inventors' reversals of fortune—the charcoal manufacturers are distraught over the illness of one partner's daughter, and Doctor Else kills his own daughter during an episode of delirium tremens produced by the orange brandy—but the narrative engine is neither sentimentality, in the first story, nor naturalist tragedy, in the other; rather, it is the unvarnished exposition of two fiascoes, as a statement about the protagonists' stubborn psychology. They are captains charting their own courses, and their delight in the routes they ply is precisely the delight of testing limited skills through ingenious practices that circumvent but never manage to bridge the gap between a venture and the essential funding and information. These inventors of the frontier, in both senses of the phrase, arose from notions of pioneering that involved technology in addition to geography. Their appeal for fiction lay in the possibilities for documenting their limitations as well as their determination to press on, with the awareness that such limitations existed as obstacles but also as a narrative and ideological driving force.

A Detour through "Science"

A detour is needed in order to move from technological invention to the type of fantastic literature known today as science fiction. A neighboring vein had been worked by Holmberg[24] and Leopoldo Lugones, in the literature preceding or contemporary with Quiroga, but something in Quiroga sets him apart. We must look at what he did with "science," from which he mined commonplaces that were already familiar to modernism, recycling them as much to explore the construction of subjectivities as to revive classical themes within modern scenarios. Although Quiroga extracted only a few literary innovations from his tour of "science," he needed it as a backdrop for his sketches of dogged but ignorant pioneers. Like them, Quiroga demanded something from science, even though, again like them, he may have had only passing acquaintance with "pure" science, knowledge divorced from practice. Science is remote; technology, proximate. For this very reason, science has an authority to which technology must finally defer. The present detour attempts to cover the distance between science and technology, making no claims to occupy some middle ground within a larger problematic: that of literary discourse and scientific discourse.

At the other end of the spectrum from naive, self-taught inventors were inventors who were university-educated, typified by those physicians who, ever since the advent of naturalism, had been bringing literature under a self-proclaimed objective eye, the eye of science. Quiroga ridiculed its objectivity, but several of his stories do contain something of this medical outlook,[25] which was the face of science at the end of the nineteenth century. It is also the case, however, that the physical and natural sciences, especially biology, were solidly esteemed as explicative schemata, particularly wherever scientific thought had been popularized—in institutes and associations, in instructional manuals, and in mass-market books.[26]

The first instance of this outlook in Quiroga refers to an etiology, quite a curious one, that is not medical but technological and is found in "Idilio," a very short text published in 1904.[27] This is a stock story made from literary leftovers, with an overarching, lightly ironic tone influenced by the decadence of late-romantic bohemia. The writing does not transcend the level of cliché; for that very reason, the technological reason for the protagonist's having gone blind, through a highly improbable set of circumstances, speaks clearly of an idea of causality that is both foreign to and distinct from the story's clichés. The individual sentence exhibits the torsion of the lexical whole's near implausibility: "Samuel went blind at the end of September—an acetyl explosion burned his eyes, extinguishing that spirited lad's sight forevermore." In the moral expounded by the story, Samuel has to go blind because, among other activities, he has been feigning blindness in order to live as a beggar. The inevitable blindness aside, however, its cause could have been a gas lamp, a blow against a window frame, the explosion of a heater, or an accident in the street, with Samuel run over by a carriage or even by one of the streetcars that have "greatly raised property values" in the new districts like the one where he lives. Instead, though, there is the "acetyl explosion," glaringly out of place; it literally comes out of nowhere. And if it does come out of nowhere (we have already learned that Samuel is unemployed), we have to wonder why it is there, and especially whether it foretells—not deliberately, but through a meaningful anticipatory causality—some of what is to come later on in Quiroga's writing.

What is it that science gives to literature? What is literature thinking of when it mentions or alludes to science? Far from a scientism of what is said, a scientism of *form*: what is said is vouched for by the form in which it is presented. Thus the recourse to science—science in its discursive form—must be indicated by quotation marks, for this is a question of the form of "literary discourse" being stamped with the seal of approval of what is thought of as the form of "science." By contrast with technology, which involves skills and description, "scientific" form proposes an explanation; accordingly, it proposes a causal scheme and—on that basis, in the literary realm—a plot. Literature does not think in the way science does, but rather in the way it *believes* science thinks; thus it makes an agreement and obtains a guarantee.

The voice of "science" frees the story from moral limits: with science comes the right to say even that which offends social convention; when it is science that speaks of a transgression, there is none. For example, a character who is a doctor has professional authorization to speak, and he is permitted to place himself outside the limits that custom imposes on the speech of others. In "Una historia inmoral," Quiroga both demonstrates and mocks this prerogative:

> "Do you think so, Doctor?" the mother stammered.
>
> The outcome of my story depended on what he would say. He shrugged, as it happens, and smiled faintly.
>
> "It's completely natural!" he said in his patronizing way.[28]

The doctor authorizes the story, and through this act he legitimates the curiosity of the audience and opens the way for a similar story, told by a second doctor to this audience that has already been prepared for scandal:

> "You probably know of many such cases, don't you, Doctor?" asked the same woman. "Though it's probably not possible for us to hear about them!"

As the title indicates, what is involved here is an "immoral story," one that combines homosexuality and incest. Here, however, it is not the story's theme but rather the narrator's success in imposing this story on his audience (and, incidentally, winning over the young woman whose attentions both doctors have been courting) that reflects the medical authorization of narrative discourse. Although Quiroga maintains an ironic distance from this authority, the story places it in a social setting, confirming its existence as an element within an ideology.

The other stories that reflect this medical outlook involve lunatics, and especially episodes in which someone literally provides the theme of a tale by losing his mind. The most perfect story in this series is certainly "El conductor del rápido," a genuine formal experiment in which an attempt is made to examine the nature of speech that falls outside the bounds of rationality, when hallucinations alternate with briefer and briefer moments of extreme lucidity in the teeth of an encroaching insanity from which only a remnant of moral consciousness can be salvaged. This is not just any insanity, moreover, but an occupational insanity linked to what was, at the time, the most advanced embodiment of modern transportation: railroad tracks, locomotives, signal systems, boiling cauldrons. The conductor of the express train is a madman whose physical extension is a machine, and whose insanity takes its strength from the power of the machine; hence the intersection of two universes, of psychiatric science and transportation technology. Quiroga sets the story within the dual framework of speed and rationality. The conductor's insanity is not just an occupational illness (or cannot altogether be seen as one); it is a delirium strengthened by the technology that it has at its disposal.

Other madmen appear in stories that are perhaps less memorable and probably more ironic—"lively chitchat, like the discussion about the insane that took place" in the home of Leopoldo Lugones days after he must have visited an asylum where "the gallantry of its population, added to what [he] had observed on [his] own from time to time, offered more than enough material for a stimulating conversation among men who were sane."[29] What were they doing, these two writers, Lugones and his friend, visiting asylums? One need not swallow these stories whole to confirm that such excursions were part and parcel of an intellectual atmosphere, and not just in the River Plate Basin. They constituted a kind of research, which other writers (Castelnuovo,[30] for example) conducted by paying visits to operating rooms or leper colonies. And this was not merely a question of following the naturalist's documentary orders, which placed value (ideologically and aesthetically) on the clinic, on diagnosis, on description, and even, at one extreme, on subjecting one's own body to experimentation.

Experimentation with drugs, for example, the consumption of which appears to have been justified by ends that were didactic at first and, later on, moral. In this sense, "El haschich"[31] is a flawless case history that painstakingly transcribes the whole range of reactions to an overdose. Narrated in a first-person voice that reveals prior experience with opium, chloroform, and ether (a virtual apprenticeship for a writer who in subsequent tales would once again plant drugs on the scene),[32] the story, using the rhetoric of the medical report, vouches for its objectivity through this form, and it vouches for its moral legitimacy through the guarantee provided by this (medical-scientific) form. The "report" and the "case history" are appropriate, acceptable genres for presenting extreme situations, genres in which Quiroga explores variations on the literary construction of a subjectivity, and moral limits concerning what literature is allowed to say.[33]

In other tales, the genre of the "report" vouches for a plot's narrative verisimilitude or referential veracity; it gives the form of "truth" to a fantastic or science-fictional premise. Stories about monkeys[34] fit perfectly into the intellectual atmosphere of an era when Charles Darwin and Ernst Haeckel (especially the latter, as we know) were read and talked about not only among the elite but also among the lecturers and public information officers affiliated with cultural institutions and public libraries.[35] These stories have the advantage of appearing scientific in that they cite arguments about man's past while bringing this past to life in a geological realm that, by Platine standards, is exotic and fantastic.

Both "Historia de Estilicón"[36] and "El mono ahorcado" enjoy the moral independence of the scientific report that is not implicated in the consequences unleashed by the actions it relates. In "El mono ahorcado," a monkey is subjected to experiments designed to determine whether it can speak, and especially whether it can conceive general ideas; to that end, it is given hashish and trained to perform precisely the task that gives it the (perhaps general) idea of suicide. The even more objectively sinister "Historia de Estilicón" tells the tale—through an eyewitness, a monkey's master, who is morally responsible for those who lose their minds because of the animal—of the monkey's erotic relations with a girl and with an old man who is quite clearly described as a homosexual. The humans end up brutalized, beaten to death; and the monkey, locked away forever deep inside the house, lives on, crushed by the memory of its deeds, humanized by its own violence.

If the allegory is clear (humans and monkeys share a biological foundation that, in certain limited circumstances, makes them equals), the distanced tone of the report and the complete irresponsibility of the narrator, who is heedless of the tragedy unleashed by his curiosity, reflect the morally unencumbered territory of positivist science. What the scientific report lends to literature is not just its ideology (as has often been said of naturalism) but also its *moral autonomy*.

A Scientific Serial Novel

But if the technical or scientific report vouches for the moral autonomy and credibility of what is being related, another genre—science fiction—presents the conflict between morals and the progress or application of knowledge. In *Caras y Caretas*, between January and February 1910, Quiroga published *El hombre artificial* in six installments.[37] This tale, with its twin roots in serial novels and works of popular science, is marked by many of the issues to which scientific-technical developments opened the literary imagination.

Its characters are straight out of a serial novel: Donisoff, a Russian prince who betrays his lineage by becoming a revolutionary and handing his friend and protector over to be killed; Sivel, an Italian doctor who sacrifices his love for a woman on the altar of scientific passion; and Ortiz, an Argentine aristocrat misunderstood by his family and disinherited. Each of these ministories might have unfolded as a serial novel in itself, but Quiroga caused them to converge on the closed world of the laboratory, where the three Faustian young men have gathered to create life from its simplest elements, synthesized by chemistry and animated by electricity. The tale also preserves the symbolic system of a serial novel, shaped by fierce opposition between the moral and intellectual realms. Donisoff, the boss, is a demonic angel, an unresolved contradiction, "cool and assured despite the great turmoil in his soul." Like a serial novel, the tale alternates between extremes of excitement and utter deflation, and it has the uncomplicated emotions of a serial novel: friendship without rivalry among the three scientists, and an extreme pain or complete anesthesia that will destroy the experimenters' lives along with those of the products or victims of their research.

But *El hombre artificial* superimposes on this hackneyed plot an assortment of "scientific" premises reflecting many of the themes that sparked discussion in the first decades of the twentieth century: the possibility of producing life artificially (a possibility connected with the issue of abiogenesis) as well as notions about the elemental constitutive substances whose combinations should yield every known life form. Here we pick up an echo of the idea, going back to Pierre-Simon de Laplace, that all knowledge about the universe can be expressed in the mapping and arrangement of the particles that make it up. Using what is known about the elements of this mapping, one should be able to create life. Electricity (the fictional power par excellence, being invisible and omnipotent) is presented as the animating principle of living beings, as energy communicated between bodies, and as a primal force. It is used to animate the body of Biógeno, the artificial man produced by the three scientists, and they also turn to hypnosis as their method of choice for transplanting experience and memory from human bodies to the body of their laboratory creation.

From this system of ideas that could have been debated at a pop-science lecture, not to mention in the popular press, Quiroga created the Platine version of a modern Prometheus: Donisoff the scientist is in effect a Doctor Frankenstein, fashioning his monstrous creation from the elemental substances of oxygen, nitrogen, and phosphates instead of stitching it together from human anatomical remains. What accounts for the difference is the century that elapsed between the publication of Mary Shelley's novel and the year 1908—between the practice of anatomical dissection, which reveals the body's mechanical structure by separating its parts, and chemistry, which isolates elemental particles in the laboratory and reconstitutes a body's invisible structure from these primal origins.[38] If Doctor Frankenstein created his monster surgically, on the dissection slab, Donisoff and his friends create theirs from elemental matter in the chemistry lab, and the human double produced in this way is more perfect, in a formal sense, because its provenance is not a patchwork of parts but rather a process that falls midway between the additive and the synthetic. The three young men are explorers and inventors of procedures, and whatever the outcome of their method may be, they discover and discuss "scientific" principles while employing it. They do not simply apply a type of knowledge; instead, they construct knowledge, but through that process their method is brought face-to-face with moral questions that cannot be answered only in terms of scientific needs, for science resists subordination to morality.

Quiroga's tale is set within this scenario of opposing principles: science, when it succeeds in creating life, touches on the realm of myth, but at that very instant it comes up against a moral limit. Can the three scientists breathe a conscience into the inert body of Biógeno, their creature, at the price of annihilating another human being by subjecting him to convulsions of physical suffering? But, more than that, can a conscience really be created from extremes of suffering, or will the whole project fail because the three scientists, even though they can reconstitute a body from its chemical elements, will never be able to construct a conscience or a system of feelings? In short, is this perfect statue, lying on the laboratory table, condemned to be inanimate because of the way its origins have defined it? The three scientists are inventors, having perfected what they take to be an unlimited technique for creating life; starting with a rat, they have gone on to build the exact simulacrum of a man. But technical perfection is no guarantee of moral perfection (as the tale shows when Donisoff, the most brilliant of the three, commits acts of the greatest immorality), nor in particular is it any guarantee of the creation of life in the sense of feelings and a conscience.

Electricity and hypnosis fail, not because it is totally impossible to connect two bodies by these means, but because (to reason by analogy) it is impossible to transplant a conscience. Positivism hits the wall in this Faustian parable[39] that reformulates the ground rules for the relationship between values and knowledge, asking yet again about the establishment of a hierarchy of values that might be able to point the way for science by clarifying which obstacles science may approach and which it must leave alone, which methods are morally legitimate, and how life, in its integrity, can be deconstructed in the material sense only, into parts whose recombination offers no assurance that new life will appear.

With Donisoff and his two friends, science runs right off the rails. First they apply their hypothesis about the fabrication of matter to the creation of life. Then they use flawed reasoning, when they regard consciousness, by analogy, as a mechanical storage battery with a genetic charge that can be replaced through the transmission of other charges accumulated in other human bodies. Finally, they do not know how to resolve the moral question posed by their experiment: Is it possible to obtain conscious life through the annihilation of another conscious life? Quiroga's serial novel builds a plot from these threads, several of which, being very old, belong to the Faustian tradition at the origins of the modern era; others, emphasized throughout the tale, spring from imagination in collision with science and with what had passed from science into the discourse of popular science, manuals, and newspapers.

The scenery and props in *El hombre artificial* are those of the laboratory as it appeared on the covers of a number of paperback novels, and in magazine illustrations (including those in *Caras y Caretas*, which undoubtedly was attuned to the "weird science" aspect of science and technology). But even as a fictional setting, the laboratory was a setting that was new to literature, and the inventor who occupied it was also a brand-new literary and social type, for he was different from the figures associated with comparatively more familiar areas of knowledge, such as the doctor in his clinic or the surgeon in his operating room. In most people's experience, laboratories and scientific inventors were uncommon; their kind of knowledge did not circulate in a symbolic universe that intersected everyday life, but in one that was radically different: it was knowledge with no immediate purpose, knowledge for knowledge's sake. The scientific inventor was the shadowy culmination of something that was also fundamental to the technological innovator, but a culmination that struggled to free itself from the social and economic goals that motivated the technologically oriented, pragmatic inventor. In this sense, the laboratory and its occupants were exotic, given most people's experience, but their exoticism could be seen as an intensification of the knowledge on which technological knowledge also depended.

By constructing his tale around these three exotic characters (who were of course cosmopolitan as well), Quiroga, a writer fascinated by knowledge of a practical kind, wrote a work of fiction in which such knowledge was projected onto the "scientific" backdrop that made it possible. He would never again return to this type of fiction, but his serial novel of 1910 marks out a zone of ideological and aesthetic contact (a serial novel in a mass-circulation magazine; fictional themes and settings based on the stuff of science) between a writer thinking about the public and a literature recycling old truisms, using themes and settings formed from sketchy versions of contemporary knowledge.

CHAPTER TWO

Arlt: Technology in the City

> Every period requires its own form. Our mission is to give a new form to our new world, using modern media. But we carry on our shoulders all the weight of our knowledge about the past: unconditional affirmation of the present presupposes a merciless negation of the past.
>
> —Hannes Meyer

The City Imagined

Theodor Adorno said that philosophy should let the big questions go. What would this mean for literature? It would mean changing not just how literature is made but literary culture, too. Which is exactly what Roberto Arlt did—he changed literary culture and fixed his sights on things that the writers who were his contemporaries could not see.[1]

Arlt made literature from found materials discovered in the modern city, from content foreign to the "official" writers, from fragments of a city that he knew better than they did, from types of knowledge that had no cachet: how to set up a brothel or smelt metals, strike gold or make money outside the dreary workaday routine, blend technical information with storytelling. Arlt said what simply *was not said* in the literature of Argentina, because Arlt, as a writer, came from a different place. Barred from the high-toned learning of the intellectual elite, he showed his resentment in the famous prologue to *Los lanzallamas*, or he made fun of symbolic dispossession in his "autobiographies." Faced with the unequal distribution of learning, he responded with a technological repertoire in all its plebeian prolificity, and he devoted his writing to what was not supposed to be uttered—here was the source of both his fiction and his transgression. Arlt was neither as revered nor as reviled as his sympathetic biographers and critics have made him out to be. He was simply talking about something else, which made him difficult, not easily domesticated. He was an alien. He had an unsettling connection with the world of the poor, based not on ideological sympathy or moral concern but on cultural territory that constituted common ground. Arlt was never able to feel that he fit in, for he felt dispossessed (and, as we know, there is no right or wrong when it comes to feelings). As one dispossessed, he walked safely where other writers, his contemporaries, feared to tread, and he looked around where they did not. He entrusted lunatics and machine enthusiasts alike with the power to reshape society and culture. This is what fed the unbelievably prescient power of his writing: in the Buenos Aires of the 1930s, he saw what would be the city of the 1940s and 1950s, and he took a leap across time, jumped over the present, because a scattering of signs in the present had become the future in concentrated form.

Arlt's imagination was obsessed with the city and technology; together, they drove him both to expand an area of subject matter and to construct a form and an ideal of beauty. The writer at large in the modern city happened upon technology; through his relationship with technology, he learned to see a city that was new to literature. The city and technology—not separate but united, in the evolution of his fiction as well as in his critical impulses. Arlt literally planned a city, for the Buenos Aires of his texts is both a portrayal and a premise. He felt at once fascinated and repelled by the city he had built in his fiction, a city onto which, prophetically, he had stamped the images of cities that were completely modern, in a montage made up of what he thought he knew about North America and of his generalizations from a handful of empirical facts about the Buenos Aires that actually existed. Arlt's way of seeing retained little of the flaneur's idleness, yielding aesthetic configurations that classified and organized images in a domain that was separate from physical space, in which the empirical city—deconstructed and reconstructed by the transformations it had been undergoing since the end of the nineteenth century—became the foundation on which an imagined city was imprinted, the city of the future, where the present would be retooled by the technical imagination.

When it came to the past, Arlt was completely without nostalgia. He did not accept the history of Buenos Aires as the city's constraining framework, much less as its outer limit. And when it came to imagining the future, the present, for him, was only a rough sketch or scenic backdrop. This historical blankness corresponds to European notions of America—our cities are young, unlike those in the weary Old World[2]—but it also reflects the indeterminate situation of new Argentines like Arlt,[3] for whom the value of the present eliminated all concern about betraying a history in which they had played no part.[4] By contrast with Borges, who in the 1920s created an urban mythology that was marked by a sense of both the historical past and the past of the city, the issue was settled for Arlt by a new literary foundation, made from the stuff of a backdrop that had been shredded by the chaos of urban growth and industrialism:

> An expanse boxed in by high facades, with a coal fog that seems to hover between them. Along the cornices, above the moldings, phosphorescent balcony doors with blue perpendiculars, yellow horizontals, violet obliques. An incandescence of gas from liquid air and high-frequency alternating current. Unlubricated yellow streetcars screeching on curves. Green buses silently causing vibrations in the asphalt of roadbeds, and in foundations. Above the verandas, a muddy soffit of grimy sky, and in the distance, orange-colored rectangles on a pitch-black background. The moon, the rim of a yellow plate, slashed by electrical wires.[5]

> They go off in silence, leaving behind silos of portland cement like a cluster of giants, cranes with arms slanting far above the factory windlasses, towers of high-voltage transformers built as insulators, with more rigging than the turret of a huge battleship. Darts of blue gas escape from the mouths of the blast furnaces, the bulge of a chain intrudes between two steel platforms, and a dull mustard sky can be glimpsed above the side streets.[6]

What Arlt saw in Buenos Aires was almost exactly what Borges failed to see. Against the pastel pinks of the early Borges he set a pure, expressionist, contrastive coloration devoid of whites; against a pleasant landscape (the Borgesian locus amoenus of sidewalks and residential districts), an iconography of open trenches and aggressive construction, trimmed out, as in Berni,[7] with an industrial collage of sheet-metal scraps and bits of cable. Where other writers, his contemporaries, saw a city in the process of disappearing, he saw a city under construction. For Arlt, Buenos Aires was yet to be: construction crews were pouring the foundations of future skyscrapers, and the disorderly facades pointed to a blend of the old, which was being demolished, and the new, which was still unfinished; the buildings' "reinforced-concrete frameworks, more beautiful than a woman," were still showing, and high above, on iron skeletons, the hammering of electric riveters could be heard.[8] He saw streets clogged with lines of automobiles, and a sky crossed by a tangle of wires and trolley cables.[9] The new city was the opposite of the city of squalor, for technological skill and devices—modern electrical paraphernalia—were liberating it from dim lighting, overpowering odors, the ugliness of poverty; the new city was erotic precisely because it rejected the traditional order and had no fear of vulgarity. Arlt, like no one else and like no one before him, saw the city in the form of a cubist collage whose chaotic beauty was an affront to moral sensibilities as much as to aesthetic organization. In the urban environment, he discovered the beauty of what was public and the beauty of what was corrupt, two themes that were already dogging the European writers, whose sensitivity to the technological revolution in construction had been growing since the middle of the nineteenth century. He discovered an anonymous, heterogeneous place where literally anything could happen:

> How fabulously indolent Corrientes Street is at night, sprawling among the great cubist buildings, with panoramas of chickens on the spit and gilded lounges, and cocaine stalls and theater lobbies. . . . At seven in the evening, this tramp of a street lights up all her bright signs. . . . Under these phantasmagoric lights, women done up like the ones in a Sirio[10] drawing pass by and set off a

> volcano of desires in the stiff-collared idlers rusting away at tables in jazz-soaked cafes. Night watchmen, newsboys, pimps, actresses, theater ushers, errand boys, scalpers, secretaries, comedians, poets, muggers, innumerable businessmen, authors, bums, theater critics, demimondaines: a cosmopolitan, alien, singular humanity rubs elbows in this floodgate of beauty and happiness. . . . And books, women, chocolates and cocaine, and green-tinged cigarettes, and murderers in hiding, all mingle in a stylized elegance that softens the harsh electric light.[11]

Downtown, the city showed off the destabilized beauty of its new buildings, and the monotonous group of little neighborhood houses was broken, little brick houses without iron frameworks or walls of glass windows. The residential district, the realm of petit bourgeois mean-spiritedness and little shopkeepers' petty hatreds, had no technological appeal or futurist vigor. To Arlt, its placid order was less attractive than the disorganized city with its haphazard juxtapositions, "buildings of ten and fifteen stories topped with little polygonal slate towers, pointed lightning rods, snaking in between isolated single-story buildings." The lower part of the city was covered with French-style rental houses and lit up by neon signs, a collage of what actually existed and of the future city prophesied by Balder, the typically Arltian protagonist of *El amor brujo*: masses of tall buildings, which in 1930 were still far from defining the skyline of Buenos Aires.[12] Downtown one also felt the horror of the old city—pestilent, squalid, chaotic, reeking of tenement fumes—a city that Arlt's characters never tire of speeding across in motor vehicles; this city was merged with a modern city more invented than observed, and pointing to what Buenos Aires would be but had not yet completely become in 1930. Superimposed on it, like a vision, was an all but imaginary city of iron and concrete, bristling with skyscrapers, more Chicago than the future Buenos Aires, whose metallic sound was the sound of jazz:

> The sharp-pointed summits of modern cities, concrete, iron, glass, momentarily trouble Erdosain's[13] repose. But he wants to escape from the prisons of concrete, iron, and glass, more highly charged than an electrical battery. Jazz bands crackle, rasping the air of the big cities with ozone.[14]

This perception of modernity as a place of high voltage and convulsive disorder, a place that—like the modernist ideal of ordered space, its aesthetic and ideological twin—has the vacuous, aesthetic regularity of a city on paper, appeals to whoever is prepared to deal with this alloy, so far removed from ideals of purity: unlike other aesthetics, Arlt's industrial aesthetic is baroque.

A Plan

As luck (which also plays its part in history) would have it, Wladimiro Acosta, a Russian architect who had emigrated to Buenos Aires after experiencing German Expressionism and designing the sets for Murnau's *Faust*, invented a structure in the late 1920s or early 1930s—the integral city block[15]—that was almost identical to a structure envisioned by Balder in *El amor brujo*:

> This type of building divides the city horizontally into two stacked zones: below, the zone of work; above, the residential zone. A man works in the lower zone and lives in the upper. A considerable amount of street traffic is replaced by vertical connections, with a resulting savings in time and energy, and the radical disappearance of the rush-hour tumult of people and vehicles.[16]

Buenos Aires, as diagnosed by Acosta, had been subjected to forces of tradition and capitalism that had gained the upper hand over (infinite) technological possibilities. In this conflict, which can certainly be viewed as classic, the anachronism of the city formed a bulwark against the reformer's vision, which at the same time was pointing to the utopian character of a profound adjustment in the framework of a capitalist society. At any rate, in the tension between a possible plan and radical social change, Acosta's thinking imposed order on the same facts of urban life that were present in Arlt: a city of apartment buildings chaotically mixed in with single-story houses, constricted streets choked by stalled automobiles, narrow canyons where pedestrians rushed from impassable sidewalks into dangerous streets—in a word, the multitudes, who at nightfall became not only anonymous but also aggressive and yet were no less the victims of the air pollution, the grimy vehicle exhaust, and the lack of light caused by an urban planning in which the sizes and placements of new houses were incoherent.[17] In short, this was a chaos in which anxiety seemed to be the inevitable companion of urban vision.

The integral city block was Wladimiro Acosta's Arltian invention: dreary, squalid disorder versus the rationalist order of a life regimented by technology and architecture. Acosta's integral city block handled activities and city traffic by opposing traditional ground-level circulation, doing so vertically or horizontally, but, either way, apart from the disorderly region of the street, as in the H-shaped skyscrapers of *El amor brujo*:

> [Balder's] plan consisted of a network of H-shaped skyscrapers; aerial cable cars could hang from their transverses.[18]

The proposal for the integral city block's cruciform skyscrapers, like the one for an H-shaped skyscraper, stood in opposition to the city that history had built. It proceeded from the possibilities opened up by new technologies, which served a futuristic function in the Buenos Aires model of economic and urban development: it was science fiction, both as novel and as urban renewal.[19] But Acosta, who had also gone to South America in search of the sun,[20] revised the basic proposal for his first integral city block: no longer would people work on the lower floors and live upstairs; instead, in the interest of better lighting, the residences would form a single block facing the north and the sun, the bar of a T whose corpus of office buildings would face south, east, and west. This plan, certainly extraordinary by Buenos Aires standards, was finally perfected in its next iteration, which wholly abandoned the historical city, where limits on transformation had been imposed by the colonial plan's blocks of buildings; the integral city block now spoke of the city of the future, not of inserting the future into the city of the present.[21]

In Acosta's vision, the skyscraper was "the extreme architectural expression of capitalism,"[22] whereas the layout of the integral city block proposed an integrative model of work with leisure, products with services, private space with public space, the built environment with nature, access to light with protection from the elements. The structure of the integral city block constituted an urban ethic, a lesson in right-minded social organization as well as the response to a demand. What the urban labyrinth suggested to literature (a place of heterogeneous confusion, revered and dreadful for that very reason, a flash point for distress and a setting for fantasy), Acosta's reforms appear to have tamed. His Buenos Aires was very similar to the one observed by Arlt; it can almost be said that they perceived it from the same streets, looking at the same jumble of low buildings and apartment houses, the same halting progress of automobiles, the same atmosphere of blocked sun into which, as in Arlt's stories, the acrid odors of urban capitalist debris could be introduced. The photographs that Acosta included in his book—high-contrast black-and-white photos with a haze of the big city's polluted atmosphere, portrayals of squalor—are literally illustrative of an Arltian Buenos Aires:

> Nauseated, he moves along the building's corridor, a vault-shaped tunnel, its walls opening onto the barred rectangles of elevators, onto doors spewing out the stink of sewage and rice powder. . . . Pausing next to the stairwell and peering at a small courtyard down below, Erdosain wonders what he has been looking for in this dreadful house devoid of sun, light, and air, silent at dawn and by night a din of female noise. . . . Down went the winding stairs, filthier than a dunghill.[23]

> He looks at the shaded patio, raises his head, and discovers farther above, making its way up the walls, a parallelogram of sky-blue porecelain embedded in the walls' filthy cement.[24]

And yet Acosta's urban renewal parted company with the aesthetic tension that took completely different pathways through Arlt's appetite for the city. Social engineering weighed heavily in Acosta's plans, in the sense of modern architecture's and urban development's shared conviction that they were reforming life and man's relationship to nature, but also in a more obsessively detailed way. Acosta may have been presenting a global idea of the city, but he was a miniaturist of urban utopias. In his integral city block, a miniature on a grand scale, the technological criteria of rationality, wholesomeness, and economy, yoked to his two obsessions of light and traffic, invariably prevailed over his aesthetic will.

For Arlt, the idea of civic beauty derived from a critique of the petite bourgeoisie's visual monotony (a basic motif in *El amor brujo*, as everyone agrees) and from cultural fantasies whose materials came from technology, cinema, and photography.[25] His imagination responded exclusively to the iconography of modernity. Therefore, his H-shaped skyscraper was a rapturously aesthetic version of new technologies, unknown to Buenos Aires:

> The ramparts of tall buildings would have to be replaced with thin walls of copper, aluminum, and glass. And then, instead of designing steel structures built to support loads of 5,000 metric tons, he would perfect the obelisk-style skyscraper, delicate and spiritual, not Carthaginian like those favored by the architects of this city without character. How did he intend to solve the problem of reflections? If he were to reply, in keeping with modern optical research, that he intended to arrange the glass so that the buildings would be pyramids, with surfaces reproducing the chromatic scale of the rainbow, the frequent response would be uproarious laughter.[26]

This science fictional setting united the materials that obsessed Arlt with the forms that he judged proper to modernity—copper, aluminum, the miracles made possible by a metallurgy owing much to alchemy and inventors' dreams—and he answered the professional engineer's routine with the artist's imagination: a city spiritualized in a material sense by a reduction in the weight of its walls, a city spiritualized by reflected light; a design in which understanding of materials was joined to the discovery of new forms, a version of design whose origins lay neither in refined taste nor in the consistency of a proposal but in the aesthetic conveyed by illustrated magazines and newspapers and by motion pictures.

Arlt's city had a kind of Gothic delirium. It was precisely when he thought he was putting up a fight against the heaviness of what he judged to be wrong with Buenos Aires that his synthesizing became baroque; it was too colorful for the understated chromatics of modern architecture. In short, his was a plebeian version of the city, both in its technological possibilities and in its aesthetic outcomes. In his vision of the city, Arlt, by contrast with Wladimiro Acosta, utterly lacked reformist utilitarianism; he was predestined for the new because he was a rootless Argentine, and the new was where his aesthetic impetus had to find its ground. He saw the future without a past, and he predicted the future in the present. His ideal of modern beauty had no room for a measure of refinement.

The Technical Stuff of Literature

> While the modern novel has been trying to determine the most delicate atomic movements in the souls of characters who have remained almost motionless in the sphere of novelistic life, modern physicists, trying to determine the architecture of the atom, also nearly motionless in a sphere without space, have been seeking material adventures whose repercussions in the physical world have had consequences that no novelist, even now, has managed to describe or attempted to write about. . . . Today's novel lacks adventure because the professional novelist, though this seems paradoxical, lacks a profession. Adventure, in the broadest human sense, means grasping the beauty of the major third at the very moment when the woman we adore is dying, or discovering how to remove an electron from a helium atom at the moment when one is kissing a girl.[27]

Arlt wrote a response to the literary culture of his time. He changed the vocabulary of literature, under the impact of images that came from news about World War I weaponry and from metallurgy, aviation, motion pictures, and popular science. These motifs offered their lexica for the construction of expressionist images. But Arlt also translated from other systems of representation. In films, and in photographs from abroad of the great metropolises, he saw Buenos Aires. In the urban landscape, and in the vocabulary of technical manuals, he found verbal riches stamped with modernity. His writing stages a response to the conflict over the languages of literature. Never before in Argentina had anyone talked this way:

> He saw faces that seemed to have passed through a laminator's rollers, and fragmented skulls, mutilated, as if trepanned by a drill.
>
> A spot lights up three-quarters of his face with an aluminumlike flash, and the apex of his cornea shines brighter than a movie star's.
>
> Suddenly the light dimmed a shade, going down more quickly than a plane hitting an air pocket.
>
> The Melancholy Ruffian has entered a zone of such intense lighting that, seen from 50 meters away, he looks like a black

> puppet hovering on the rim of a crucible. Industrial-gas signs slither up the colonnades of buildings. Pipes filled with yellow gases, fastened between frameworks of red steel. Advertisements of methylene blue, green rays of copper sulfate. Prodigiously tall cranes, black chains of winches turning on pulleys and oiled with hunks of yellow grease.[28]

When the Astrologer lists the potential members of a secret revolutionary society—"those who have a plan to save the world, clerks who hope to become millionaires, failed inventors, people who have been laid off or fired, those who have just been sued"—he outlines a world quite familiar to Arlt. In this socially heterogeneous enumeration there are features of a common culture, the culture of those who are aliens in the realm of the lettered elite and the good citizens. Criticism has already had much to say about how Arlt's novels can be linked with serial novels and bad translations. But Arlt can also be understood as having been fascinated by new texts and practices in technology and science, in chemistry and physics, and in the popular-science simulacra that were circulating in the Buenos Aires of the period under the labels of hypnotism, Spiritism, parapsychology, and telepathic transmission. Neither Arlt's writing nor the desires of his characters can be understood without reference to knowledge gleaned from newspapers, magazines, and inexpensive manuals as well as from the public libraries that served every working-class neighborhood and from the workshops of crazy inventors hooked on electricity, fusion of metals, galvanization, and magnetism. This was poor people's knowledge, an assortment of discourses occupying the same place in the education of indigent urban intellectuals that other types of knowledge had among the social elites. It was practical knowledge doing double duty—as a myth of upward social mobility, and as compensation for poverty of symbolic capital and insecurity over scholarly capital. Poor people's knowledge was a sketchy, information-oriented version of advances in science, especially chemistry and physics, although quasi-fantastic discourses could also be found on health and illness, miracle cures, subjectivity, normality, and insanity.

Fantasies Old and Modern

The first known text by Roberto Arlt, "Las ciencias ocultas en la ciudad de Buenos Aires," evokes an underworld of parapsychologists and Spiritists, a world that often intersected the world of working-class followers of new developments that were presented as scientific (in fact, Erdosain, who is obsessed with technology, meets the Astrologer at a Spiritist center).[29] "Las ciencias ocultas" is an ideological protonovel in which Arlt deals for the first time with contact between types of knowledge: the pasty young man who draws the narrator into the theosophical society meets him in a bookstore where he has been looking for a history of mathematics; the foundations of the occult sciences blend archaic origins with "the modern phenomena of hypnotism, magnetism, Spiritism, and radioactivity"; refutations of these sciences' theories are carried out with elements of a science that has been learned from books about chemistry, geology, and biology, and from which fantastic ideas have been culled about the nature of the sun, the cosmos, and the structure of the universe. Thus the refutations partake in the imaginary character of the discourses they seek to invalidate.

Nevertheless, theosophical discourse made it possible for Arlt, an indigent young man, to take a stand from which he could dispute one type of knowledge from the perspective of other types. In this sense, "Las ciencias ocultas" is a formal exercise in debating, in which Arlt, at the age of twenty, constructed a strategy of resistance to the twin forces, knowledge and power, that would engage him for the rest of his life. More important than this text's exposure of theosophical fraud is its self-imposed exercise of argumentation, with reasoning, footnotes, and verification of sources. Arlt learned to argue and use books, whether in constructing a proof or in assembling the portrayal of a hypothetical world. One needs some type of knowledge in order to create literature; and, in particular, one needs to know how a type of knowledge is expounded, transformed, or betrayed. In "Las ciencias ocultas" Arlt found his proving ground, for that is where he took possession of unfamiliar discourses and learned how to use them. He was excluded from other initiatory rites, but those of the theosophical group nevertheless offered the model of a closed society; it was able to serve as the setting for an intellectual education that clearly made Arlt's initiation different from that of every other contemporary writer except Castelnuovo.

But "Las ciencias ocultas" also speaks of a question that the Astrologer raises in *Los siete locos*, and that Erdosain, in *Los lanzallamas*, exhibits as a psychological and moral symptom: What can be done with a world that the gods have abandoned? How is it possible to offer disenchanted societies, which have lost the sense of sacredness and mystery, a mythology that can forestall disintegration and afford freedom from a routine that, because it has no ultimate objectives, is simply barbaric or insipidly repetitious? Men and women were seeking new gods in the doctrines of Annie Besant, Madame Helena Blavatsky, and Allan Kardec,[30] but theosophical fantasy could be vanquished by still other kinds of modern illusions:

> The Astrologer is right, then. The day is coming when the people will make revolution, for they have no God. Men will go on strike until God shows himself.[31]

> And, by thinking, I reached the conclusion that this was the dreadful metaphysical sickness in every man. Humanity's happiness can rest on nothing but a metaphysical lie. . . . Deprived of this lie, humanity falls back on illusions of an economic nature . . . and then I remembered that the only ones who could restore humanity's lost paradise were flesh-and-blood gods: Rockefeller, Morgan, Ford. . . . The time is close at hand when skeptical humanity, driven mad with pleasure, made blasphemous by impotence, will become so enraged that it will have to be put down like a rabid dog. . . . The pruning of the human tree will be . . . a vintage that millionaires alone will be able to perform, with science at their service. . . . This society will comprise two castes, and between them there will be a gap—or, better, an intellectual difference—of thirty centuries. The majority will be scrupulously kept in the most absolute ignorance, surrounded by miracles that, being apocryphal, will be far more interesting than historical miracles, and the minority will be the absolute repository of science and power.[32]

The question of the social order, a question that runs through *Los siete locos* and *Los lanzallamas*, has two answers, one technological and the other mythological. In Arlt's novels, these two dimensions are united—technology, allied with wealth, makes it possible to produce both a social order and a mythology, provided that technology is managed by a body of experts: true, utterly fearless revolutionaries who exploit the resources of technology and science to erect the foundations of a new political order. Industrialism is not just wishful Bolshevik thinking, not just the engine of Western capitalism or expansionist militarism; it is also a literary form of this modern fantasy: "industrial mysticism" rests on an aesthetic transmission of science, through whose resources a pantheon of "supercivilized gods" becomes available.

It comes as no surprise that, in the late 1920s and the early 1930s, a character in a novel should have discovered the disenchanted void and envisioned the authoritarian restoration of a unified society. In other words, such fantasies were the order of the day, and it was not unusual to see intellectual reactions to the mass democratic and communist movements, with their dreams of a more strongly consolidated society. In this sense, Arlt invented nothing new, nor was there anything new about a pessimistic utopia in a world where science and authoritarianism functioned as political allies.[33] It is striking, however, that these reactions were taking place in the literature of a Latin American country, and that both of Arlt's novels take a singularly ambiguous view of the scientific revolution.

The instrument of a society that was authoritarian (and deranged by its authoritarianism) was at the same time the stuff of fantasy and a source of beauty. Arlt wrote one chapter of the industrial aesthetic, which had lethal possibilities but produced poetic identifications and fantasies. Erdosain dreams a technological dream in which destruction and beauty are interlaced in an ambiguous relation of necessity. And even though a form of madness might be seen in this dream, it is impossible not to see as well a category of fiction free from moral or psychological judgment. The industrial landscape, which heaps metal, rails, and cables onto all that remains of nature, is beautiful; the invention of total destructive power produces the effect of a cinematic backdrop as appealing as it is appalling. This dream is monstrous in its radicalism and lovely in its portrayal, the aestheticization of a setting that anticipates Hiroshima:

> He seems to be in an industrial area. Rails trace arcs alongside mountains of coal and black oil tanks. Locomotives with bronze fittings and conical smokestacks maneuver along the gauges, pushing cars piled high with rocks. The bridges groan ferruginously under the speed of the passing express trains, and the slaves go in and out of their soot-black quarters.... Suddenly, on a tower platform next to him, a glass torpedo lights up, greenish, like a Crookes tube.[34] The atmosphere is charged with static, and suddenly a rectilinear, conical discharge of green light causes the porcelain corkscrews to explode. A locomotive rears up on its wheels, hesitates a fraction of a second, and explodes into three distant points of molten metal. Erdosain, in his leaded-glass control booth, smoothly swings the glass torpedo about. Rays strike against rocks, and the foundations of dwellings explode. He even notices the following detail: in contact with the rays, a woman's hair stands on end while her body is reduced to ash.[35]

Technology made its mark on the Arltian text. The "diabolical formulas"[36] introduced a strange system of symbols whose visual purpose was to make the page strut its "scientific" stuff even before it was read. Likewise, the sheaf of papers that Erdosain hands to the Astrologer, containing the plans for building the phosgene factory, is incorporated into the text of *Los lanzallamas* not as a digression spinning off from the tale but as a destination at which the tale must arrive: the diagrams, explanations, formulas, and explanatory subtitles are at the center of the narrative, not at its margins. They are indispensable if the Astrologer's dreams of a scientific revolution, and Erdosain's dreams about the need to invent "supercivilized gods," are to show the stuff of which they are made. These dreams have sprung up from the territory undergirding modern mythologies, because men, abandoned by the gods, seek an idol to take their place; the point here is to show technology, not just name it; the name alone might appear to mention technology as if it were something going on somewhere else, whereas the grapheme posits technology as the body of the text. And the name is never sufficiently real, whereas visual representation demands that technology be recognized even if it is not understood: here is the machine, and here are the formulas for making it work; here is a kind of knowledge, not just the name of a kind of knowledge, not just a reference to a kind of knowledge. The diagram simultaneously authorizes and completes an industrial aesthetic, in stories filled with dialogues about technology and with dreams made from the stuff of technology.[37]

The revolution the Astrologer has in mind is politically heterogeneous in content and is based in particular on a bold, interventionist style of seizing power, one that evokes the audacity of the fascist putsch, the White Russian group of counterrevolutionaries, Sorelian[38] voluntarism, and Leninist centralization.[39] The power that is sought has no fundamental value in itself; it is posited as an authoritarian, organizational response of the body politic—this revolutionary end is of less interest than the means of its attainment. The means are central, and for this very reason Arlt created a fiction whose originality consisted in his constantly blending the discourses of psychology and morality with those of technology. The two dimensions created an interdependent relationship: madness and dreams, like an anticipatory setting and the dystopia of the postrevolutionary world, were about "technology."

In this respect, *Los siete locos*, and especially *Los lanzallamas*, are science fiction stories in which a society of the future is not so much portrayed as conjured. These two novels, instead of depicting and describing that society in the present, depict and describe its potential: if such and such conditions are fulfilled (if power is seized by technorevolutionary means), then the society of the future will have such and such organization, based on a technocracy that will impose its power through repression, but also through the cohesion supplied by a social imaginary that operates with the resources and contents of technology. The problem that confronts the Astrologer in his delirium is a simple one: How might knowledge be used to change relationships of power? By training "revolution specialists" in camps where life would be ordered by advanced technologies, and where the cadres would be given technical knowledge. The only things needed for revolution would be the will of a hundred trained men and the deadly power of phosgene—insurrection plus martial industrialism.

The spectacle of this revolution was to unfold within a science fiction scenario taken up with machinery and weapons that received wide exposure in newspapers and magazines after the experience of World War I, and that were harbingers of the future more than chroniclers of current conditions. Chemical and biological weapons, gas masks, and enormous cannons had been emblems of the future in World War I, which was still using the cavalry and trench warfare of earlier times. But gas masks and nerve gas were the most dreaded and the most strongly evocative images of an event that had rapidly ushered the technological promise of the future into the present. This technological aspect of the war did not disappear from newspapers in the 1920s, and it was supplemented by a proliferation of inventions involving long-distance communication, new materials, and new industrial technologies.

The science fictional aspect of *Los siete locos* and *Los lanzallamas* is rooted in actual events, and in speculation about what was to come. The characters in these two novels use current knowledge to imagine future organizations; to questions about society and politics, they respond in terms of technology. They are delirious with technology, and their delirium is also a predictive science fictional premise.[40]

The Novel of the Indigent Young Man

> The commissioner of Patents for Inventions and Trademarks, Business, and Agriculture hereby Certifies that, in recognition of the formalities established by the Law of 11 October 1864, the following decision has been handed down . . . in favor of Roberto Arlt.

On October 17, 1934, Roberto Arlt was granted the patent for invention number 42.050, "stockings with toes and heels reinforced with rubber or its derivatives."[41] He was already a renowned writer, with successes in the theater, a high-profile job in journalism, and an established name as a novelist. Why, then, in 1934, this loyalty to technology, that modern form of dreaming for legions of amateur inventors?[42] A psychological answer to this question might prove both obvious and inaccurate; perhaps an explanation can be sought in Arlt's fiction. Technological fantasizing was a mode of dreaming for successful inventors, just as it was for poor men in search of recognition—for example, Silvio Astier in *El juguete rabioso*,[43] or Erdosain and the Espila family in *Los siete locos* and *Los lanzallamas*.

The exaltation of the inventor, which appears strange in a programmatically hopeless, pessimistic literature, was commonplace. As in a Frank Capra movie, this modern dream transported the inventor from poverty and obscurity to riches and fame. Invention, "just like in the movies," was practically the only road that had not been walled off from the beginning; on the contrary, a change in social status entailing upward mobility, and using the imagination to resolve conflicts provoked by inequitable distributions of wealth and knowledge, occurred through a stroke of fortune, or through invention—or, perhaps better, through invention *as* a stroke of fortune, the formula for a plebeian dream derived from North American fantasies (that is, from the legend of a successful inventor like Edison, and from cinematic plots). One might say that a pair of rubberized stockings or a metallized rose was a minor discovery for technological fantasy, which (to move from science fiction to scientific parody to deft modifications of machinery in everyday life) also encompassed the death ray, a dry-cleaning plant for dogs, and a simple automatic transmission for a steam engine.

Silvio Astier and Erdosain present themselves as inventors, a class of intellectuals who are nevertheless uncritical consumers of contemporary journalism, and who assemble themselves into institutes, circles, and clubs. The Astrologer, answering the call to this identity, supervises the tasks of a secret society in which labor is divided according to carefully differentiated specialties. Erdosain, for his part, refuses to work, since he considers himself an inventor, and he persuades the members of the Espila family to devote themselves to the development of his ideas for coating flowers with metal through a process of galvanization. He steals in order to outfit the workshop in the boondocks of Ramos Mejía where his invention is to be manufactured, and he convinces the Espila siblings that all of them are going to strike it rich.

The five Espilas are shown in the glow of an acetylene lamp:

one of them, perusing a "review of hieroglyphics," knows differential and integral calculus; Elena is studying galvanization; and Luciana, who is in love with Erdosain, rhapsodizes about her love in a language that features blast furnaces, Bessemer converters, and the irregular or perfect separation of carbon in the manufacture of steel. In the wooden shack in Ramos Mejía, the Espilas, like Erdosain, are consumed with passion for a fabricated rose, which they believe they can manufacture on an industrial scale:

> Luciana, impatient, leapt up and pulled open a drawer in the washstand, and Erdosain smiled, excited. The blonde young lady held the copper rose upright between her fingers, and the marvelous metallic flower unfurled its reddish petals in that wretched shack. The acetylene lamp's trembling flame projected a play of transparent red, as if the flower were being animated by a botanical life that, having been consumed by acids, now constituted its soul....
>
> Remo examined the copper rose once more, admiring its perfection. Each of its red petals was nearly transparent, and the natural petal's veined form could barely be made out through the metal coating, for the glue had darkened. The flower was almost weightless....
>
> Lifting his eyes from the rose, he met Luciana's gaze. The young lady's eyes had a mysterious velvety warmth, and her smiling lips allowed a glimpse of teeth that gleamed.[44]

Creativity in applying practical kinds of knowledge is the aesthetic of the poor. Before inventing the death ray and phosgene, which depend on procedures for chemical sublimation, Erdosain comes up with a kitsch item whose facile, sham beauty can be incorporated into the banal decor of everyday life. This most literary of flowers is able to withstand the most corrosive processes of metallurgy; it can also bring poetic illumination to the Espilas' shack while answering to a tawdry dream of wealth. Although it is described in the same way in which modernism has described works of art, it is the product of treatments performed in a makeshift metallurgical plant. The rose is displayed both as a product of technology and as an allegory: through galvanization, the flower overcomes its own fleeting temporality, which has also sustained its poetic force. Losing its evanescence, it loses its aura; in exchange, it acquires the petrified permanence of the metals: technology has modified its allegorical power, transported it somewhere else.

The metallic flower is an oxymoron. It presents a contradiction in incompatible, inseparable terms, thus revealing the antireferential possibilities of the poetic imagination, whose impetus in this case is a technology-oriented fantasizing that has managed—miraculously, and in advance of encounters with actual machines—to think up this hideous object: a flower forced into a metal casing, scorched by an excess of electrical current, and darkened by cyanide that was converted to copper cyanite and corroded by a wash of nickel during the process following galvanization.[45] To achieve the oxymoron of a copper rose, the Espilas have given the flower a coat of glue that has darkened its petals, but it still has the look of being alive, if blackened and burned by its acid bath. At any rate, it looks splendid, and in the perverse light of the acetylene lamp it glows a deep red, almost as perverse as the rubber-lined orthopedic stockings that Arlt patented in 1934.

Erdosain follows a trajectory that—notwithstanding perverse inventions and pointless ones, the machinery of war or vulgar ornaments—encompasses many of the era's technological fantasies; his specialty, if it had to be named, would be metallurgy and chemistry, but he knows something about everything, and every one of his new technological obsessions, from movies to airplanes, radio, and energy production through atomic fusion, has its place in his speech.[46] When Arlt sought the patent that he obtained in 1934, he was one among hundreds of amateurs who had managed, in heated competition with the expertise of local and especially foreign industry, to make their way through the bureaucratic labyrinth and find, in patents that never saw commercial use, public declarations of legitimacy for laborious procedures attempted with machinery that was insufficient for the technical requirements as well as for the freedom demanded by the inventor's imagination.

Arlt was the cutting edge of something pervasive but vague that was taking place in parts of society somewhat removed from traditional, modernizing, or avant-garde centers of cultural initiative. Using scraps of throwaway knowledge, he put together an unprecedented theory about the place of modernity in Argentine culture. That place was a conflictive and vexing one because intellectuals had not been thinking about the technological aspect of modernity. But that is precisely what Arlt had been thinking about, which is why his imagination seems so strange and incongruous compared to that of any of his highbrow literary contemporaries. He inaugurated a technological utopia that, like any other, was a plan for radical change in actual conditions but, above all, a diagnosis of the reality postulated and constructed by speech. The Arltian character speaks with the passion and hope of an underclass utopian—the Astrologer's society of the future is no less dreadful than other twentieth-century utopias. Arlt shares his characters' radicalism and their loathing of compromise, but his greatness lies in his having looked in a new direction, both to create his fiction and to find the material of its creation.

Arlt's great do-it-yourself project, like those of the amateur inventors, is marked by an aesthetics of amalgamation; above all, from beginning to end, it shows what a common man can create with writing: a literature of the dispossessed, full of resentment, ambition, rage, greed, and hardship.

CHAPTER THREE

Popular Science and the Popularizing Press

> Within a hundred years, every citizen will have a helicopter made by Ford.
>
> —*Crítica*

Newspapers and Magazines

In 1925, when Albert Einstein visited Buenos Aires, *Crítica*, the city's most modern daily newspaper of the period, and its most popular, devoted nine articles to him that culminated in an exclusive interview.[1] Four years later, both *Crítica* and the new daily *El Mundo* were continuing to cover the theory of relativity and gather news about the debates and refutations that pertained to it. The topic was important enough for *Crítica* to have purchased the international rights to a series it touted as "valuable early findings": articles about the theory of relativity that made a hash of terminology but a smashing impression of up-to-date, sophisticated science.[2] At the other extreme of editorial offerings, aimed at amateurs and handymen eager to improve their technical chops, both papers maintained and updated regular sections devoted to radio and automotive technology.[3] And there was practically everything in between. But this thematic variety, if curious in itself, was no more so than the number of articles and the growing volume of space devoted to topics in science, engineering, medicine, parapsychology, faith healing, clairvoyance, and wonders of nature. The year 1923 had marked a turning point, when *Crítica*'s sporadic attention to these subjects became systematic. *El Mundo*, for its part, had from its beginnings in 1928 been putting almost the same emphasis as *Crítica* did on science and engineering, but it had been doing so in another style, one that differed both in tone and in choice of subject matter.[4]

Together, the two papers also constituted a hands-on demonstration of the processes involved in the technologizing of newspaper production. *El Mundo*, the first Argentine daily to be published in tabloid format, was deliberate in its adoption and stabilization of the recent changes in journalism, from its use of diagrams to the writing and editing of its articles, from its decisions about sections, headlines, and graphics to the status and prerogatives of its reporters.[5] *Crítica* had been carrying out a content- and technology-modernization program ever since 1921; once it was sure of its mass audience, it acquired a building of its own and began to compete with *La Nación* in the areas of printing and Linotype equipment. *Crítica* adopted photoengraving a year before *La Nación* did, and in 1927 the newspaper moved into its building on the Avenida de Mayo, with a Hoe rotary printing press that turned out 160,000 copies per hour:[6]

> The rotary presses, as tall as a transatlantic liner, fill the plant with the deafening noise of the sea crashing against a breakwater. The dizzying slippage of sheets of paper between black rollers. The smell of ink and grease. Men walk by, reeking of sulfuric acid. The door of the photoengraving studio was left open, and flashes of purplish light are spilling out.[7]

High technology was carving out both a concrete place and a virtual myth in the organization presided over by Natalio Félix Botana; there were frequent articles about *Crítica*'s physical production, with topics that reinforced the paper's strategic modernity at the discursive and generic levels, and these topics were backed by assertions of *Crítica*'s technological modernity. The "hundred thousand copies per hour" became material for a journalistic saga[8] that had the Hoe rotary press as its deus ex machina. The Hoe press, more than just a technological resource, was also an allegory—abundantly represented in the paper's graphics—of modern mass journalism.[9] To reinforce this discourse, in 1925 *Crítica* (the "first in the world" to do so) founded a radio station, LOR, which broadcast from the Teatro Coliseo. After 1922, *Crítica* joined the international communications circuit when it contracted with a wire service (until then, only *La Nación* had used one), adding the International News Service in 1923 and the Associated Press in 1925.

Crítica and *El Mundo* shared a cultural arena to which each brought a different editorial outlook and journalistic ethic. Sensationalism was a stylistic feature and marketing strategy at *Crítica*, where the news was literally constructed: bits and pieces of information were pulled together, topical series were established, and wire-service stories found a response in local discussions of their particulars, through a counterpoint of reporting, letters from important readers, and polemics. At *Crítica*, the use of graphics—photographs, diagrams, drawings, and cartoons—was just short of lavish, and the paper saved the front page, as needed, for news about happenings in science or technology. At *El Mundo*, the style was closer to what would be recognized today as that of a "serious" paper, and *El Mundo* was not so patently devoted to speculations about news of the nation and the world. Stories were often printed right off the wire, without further comment from the editors, thus presenting material that was not so compelling from a fictional standpoint.

But visitors to the editorial offices of *El Mundo* did find, as they also found at *Crítica*, reporters who were willing to write articles about their obsessions—their inventions, their theories, their experiences at the outer limits of parapsychology and occultism, their discoveries, and the like. At both papers, but to a different extent at each, technical information, how-to articles, and the educational mission of the special sections were combined with columnists' opinions and their enlargements on items that had been pulled off the wire. If *El Mundo* had discovered in Roberto Arlt a master of fictionalized news and subjective commentary, *Crítica* could strike the very same note whenever one of its reporters sat down to punch up a wire-service story or lend an ear to some esteemed or quaint visitor who had turned up at the paper's editorial office, no doubt having noticed the commercial and professional sympathy with which the paper reported on the unusual, the exceptional, and the unexpected. *Crítica* was concerned less with substantiating news stories than with gauging their impact; *El Mundo*, without falling into a pompous editorial puritanism, reported the news in ways that lowered the risk of its having to correct information that had already found its way into print. *El Mundo* was more cautious but also less creative.

At any rate, both papers worked with similar material—local inventors,[10] international inventors and inventions, medical and biological experiments, the synthetic production of new substances, aviation and other revolutionary forms of transportation, long-distance communication, film and television technology, and international medical developments that had local repercussions (such as the cases of Serge Voronoff and Fernando Asuero).[11] The two papers also shared a devotion to Thomas Edison and an admiration for Henry Ford (a pair of American-style heroes to whom even the Bolsheviks, like the rest of the world, were not indifferent) as well as a respectful interest in Einstein and in following anything that Guglielmo Marconi might be up to. The traditional figure of the scientist-saint, dedicated to the future outcomes of his research, was joined by the new and much more brightly illuminated figure of the inventor-entrepreneur, who was living out his intellectual vocation and his talent in the present. *Crítica* placed more emphasis than *El Mundo* did on two areas: predictions concerning the future, the city of the future, robots, interplanetary travel, and the cutting edge of progress; and the hodgepodge of news—about conjoined twins, animals, two-headed or headless little girls and boys, congenital or pathological deformities, and assorted freaks and geniuses—that had captivated readers in the final decades of the nineteenth century and the first decades of the twentieth, and which continued to have strong appeal.[12]

For the same reason, *Crítica*, more consistently than *El Mundo*, tracked news about medical miracles, unorthodox methods of healing, and biological experiments as well as about lung grafts and grafts of other organs or glands, extracorporeal blood circulation, new anesthetics, cures brought about by rays, machines for treating insomnia, and genetic engineering; all of it fueled a journalism that had no fear of needing to issue corrections. Indeed, there was no place at all for corrections at *Crítica*, which might shift its editorial stance from high praise to vilification in the course of a few months or, if it had to, mere weeks.[13] At one and the same time, *Crítica* might be on a crusade to popularize faith healing or debunk fake clairvoyance, and such a crusade would coexist with respectful treatment of Spiritism and parapsychology. This blend of credulity, sensationalism, deception, and debunking said two things about *Crítica*'s readers: on the one hand, it reflected their belief systems; on the other, it reproduced a motley intellectual environment where the latest developments in engineering coexisted with the archaic fantasies that this kind of information was recycling in a new language, and with new possibilities for writing fiction.

At issue, in any event, was the proper response to what was also an editorial problem: how to present news about science and engineering, which had begun to be a regular offering. *Crítica* was the first newspaper to face this issue more or less head-on, for two reasons: because of its ideology of commitment to modernity, and because of the type of audience it hoped to attract, that is, readers who were more attuned to the new than to the traditions of literate culture. Sections specializing in popular science or technical skills were only one response, and from an editorial standpoint it cannot be considered to have been more than partial. Such sections fell into the category of customer service, since readers were assiduously courted by the mass-circulation dailies, but the special sections neither influenced the parent publication's style nor allowed themselves to be influenced by it, being too technical for stylistic questions even to come into play. Special sections could increase a newspaper's audience or help diversify its reader base, but they did not set the paper's style.

But they were still significant, for they accommodated at least two types of reading and two types of readers. One group of readers had a practical interest in technical matters, and they looked to the paper (or to magazines) for answers to the questions that might arise in a home workshop or at a handyman's workbench. These readers wrote letters to the editor and to the paper's letters columnist, and sometimes they dropped by the newsroom with their innovations or discoveries. They were familiar with the language of the special section, and for them its diagrams were not so much icons of modern technology as schematics for the tools of a trade. The special section might attract new readers or spark new interest in technical subjects, but its bread and butter was this first bloc of readers, whose interest in these kinds of subject matter antedated the section that had such topics as its central reason for being. These readers had to be given precise instructions, and their intellectual potential could not be underestimated. The problem they posed for the daily paper—by definition a generalist, culturally universal mass medium—was this: How could the newspaper present specifics, without watering them down, to readers who were in the know but still avoid content that would be unintelligible to the rest? The special section could also appeal to this second group of readers, at least in theory: it acted on them iconically, creating an effect of technological modernity without requiring them to read or understand its content word for word. These readers were not such attentive consumers of the special section; they might glance at it while turning the page, linger over it out of curiosity about the unknown, or become accustomed to the look of its graphics (since the special section appeared at regular intervals), or they might even read it by following the ideological-aesthetic outline of its hieroglyphics, in the same way that one can watch a game of chess without understanding it. And yet the iconic power of the special section was translated into knowledge for this second group of readers, too, just as it was for the first, even though for the second group this knowledge was not technical in nature. The special section served the purposes of cultural modernization: a new lexicon was being learned, and it corresponded to objects or technologies such as radio, which was rapidly being incorporated into everyday life, and television, whose future place in everyday life was more or less assured. This was more than just a utilitarian kind of learning; it involved a Bovaryesque pleasure in technical information that reflected an unknown but enticing world whose rules were unfamiliar but whose stock was on the rise, and whose content was not understood even though its form and face could be powerfully productive of fantasy.

The special sections had this capacity to generate myth as well as praxis because the main body of the paper was also on the lookout for specialized subject matter and viewpoints. Bits and pieces from these sections appeared here and there throughout the rest of the paper in the form of new topics, thus providing in the newspaper proper both an anchor and a preview of the special sections. Moreover, these sections came and went and were constantly undergoing reorganization; changes and improvements in them were announced frequently, as if the paper's executives and its audience could tacitly share in the simultaneously instrumental and experimental nature of such sections.

Therefore, if *Crítica* and, later, *El Mundo* could be described as journalistic operations concerned with reaching a mass audience whose varied interests were validated by special sections, it is also true that neither paper relied solely on these sections for its journalistic profile, which was based on a veritable discursive summa, ranging stylistically from the police blotter to current political events. In order for subject matter dealing with innovations in science and technology to have any journalistic credibility, it was also necessary for the other, traditional subject matter of print media to change by becoming differentiated from the rest of contemporary journalism as well as from the newspapers that had been founded before 1920. Successful expansion of the topics covered by the paper required a global change in journalistic style, and especially, when it came to the rest of the Buenos Aires press, a strong ethic of differentiation on the part of the newer papers (which had new readers). Columnists like Roberto Arlt were virtual trademarks of the papers that published them, and yet such writers could not have produced the journalistic revolution of the 1920s all by themselves. What mattered was to have a strategy incorporating all these elements but also specifying a characteristic style for pages that lacked the byline of a star journalist, or the kinds of breakthroughs in subject matter that could be used to create new special sections.

If these newspapers actually did manage to tune in to a cultural idea and become a hit with the public, it was also because they gave rise to that public, whose own social and demographic base had not been enough to bring it into being. In order to do so, the newspapers had needed to discover widespread but amorphous concerns and give them an immediately recognizable, systematic journalistic form, one that was reproducible in terms of how the paper was written and how it was read. It meant nothing that four articles on technical subjects had appeared in *Crítica* in 1916; by 1924, the appearance of one such article per week had become the rule for a mass-circulation newspaper's format. Readers knew beforehand what they were going to find in the paper; in turn, however, their advance knowledge inspired their interest and loyalty, and it was in this closed loop that the secret of journalistic success was found, when such success involved the cultural renovation of a collective entity—the public.

Both *Crítica* and *El Mundo* were published and read in an environment where tens of thousands of copies of other publications were also being sold.[14] The journalism of popular mechanics had been born. Five magazines in this genre appeared in 1922; there were seven in 1923, six in 1924, three in 1925, and six in 1926, and these continued to be published in the years that followed. At the end of the decade, in the kiosks of Buenos Aires (and in scores of towns throughout the country, to judge from the letters columns) it was possible at any one time to buy no fewer than seven magazines devoted to radio, to photographic and cinematic technology, and to technology-based hobbies, popular science, and amateur invention.[15] It is difficult to reconstruct the impact of these magazines, in the quantitative sense of how many thousands of readers they had; nevertheless, there are a few facts that appear to be somewhat independent of the magazines' own records and statements to advertisers.

Ciencia Popular, the most successful experiment of this kind, was launched in 1928 by the same press that had been publishing *Radio Revista* since 1922, also with notable success. *Ciencia Popular* was a monthly that took hold very quickly and offered the clear contours of professional journalism, by contrast with publications like the one from which it had emerged. Rather than a hodgepodge of articles that won readers' loyalty despite having been thrown together with no real graphic or editorial standards, *Ciencia Popular* was a project to which *Radio Revista* assigned its best and brightest: Jorge Duclout, the most famous of the ham radio operators, a lecturer, experimenter, and designer of radio circuits, and the author of widely read technical manuals.[16] Just a few items of information will suffice to show how *Ciencia Popular* was strengthened under Duclout's editorship. The first issue, published in August 1928, had fifty-four pages, ten of them devoted to advertisements; issue 13, from August 1929, sold for the same price (thirty cents, or less than the cost of a pack of cigarettes) and had seventy-four pages, with twenty-five pages of advertisements compactly gathered into an opening section of twelve pages, a final section of five pages, and an eight-page insert in the body of the magazine, which also included an important section of classified ads.

What did popular science mean to *Ciencia Popular*? In the first place, it was science that was basically technology, not only in its presentation of new procedures—which were quite far removed from the world to which amateurs had access—but also and especially in its practical, how-to explanations. The magazine defined its audience as encompassing the world of the do-it-yourselfer as well as that of the modern artisan; their two types of activity had much in common, given their shared desire for learning, and for research into new areas in which trained professionals were not readily distinguishable from handymen. Therefore, *Ciencia Popular* was didactic in its emphasis on illustrations, diagrams, drawings, explanatory sidebars, summaries, reiterations, answers to readers' questions, advice, comments on experiments that had been performed and encouragement for those that were just starting, prizes for handymen who had proposed simple and useful solutions, exhortations to perseverance, and contests based on ideas for advances in homespun bricolage or in efficient use of the gadgets available to handymen and proprietors of one-man workshops.[17]

At the same time, *Ciencia Popular* knew that many of its readers, by virtue of their interests, training, and activities, belonged to a world fairly similar to that of the people who wrote for the magazine. The magazine's correspondence with its readers was filled with their suggestions, which might be turned into articles; in many issues, a recurring short paragraph urged inventors to come forward with their projects:

> Have you made an interesting discovery? Found a practical solution to one of the thousand little problems we face every day? Write to us. Make yourself useful to your peers—tell them about your ideas in this magazine. Prizes awarded for published contributions.[18]

These readers (and sometime contributors) were creative do-it-yourselfers, and they were demanding judges of the technical solutions proposed by the magazine. They were the inspiration behind articles explaining how to turn a Ford into a race car (within a few years, touring-car racing, in which competitors made use of chassis, engines, and bodies from Ford and Chevrolet coupés that had been souped up in small machine shops, would catch on as a special sport of Argentine origin). They were also the audience for advice columns featuring letters whose writers seemed to have come straight from experiments conducted in semiprofessional or home workshops.

But the magazine was more than a forum for mutual recognition, or for conveying the kind of knowledge that could have almost instant application. There was also—along with photography, movies, and radio—the future promise of television, which fascinated Duclout; indeed, *Ciencia Popular* once presented a televisual experiment conducted on a TV set that the editor himself had built.[19] This is what gave the magazine its place at the forefront of the almost unbelievable experimentation in the Buenos Aires of the 1920s, but also at the heart of a subject to which the mass-circulation newspapers were in thrall. Television was a familiar technology projected into a dimension that was not yet accessible; it was the technofictional element that lit up the everyday practice of the hobbyist, the technological dimension that foretold the future and, as such, constituted one of a body of innovations whose possibilities were on display or up for discussion. And, apart from television, there were other portents of everyday life in the future—the interplanetary rocket, the artificial creation of life, the "rocketmobile," passenger aviation, seaplanes and parachutes, automated garages, submarines. In any event, the primary emphasis of *Ciencia Popular* was not on these wonders but on the types of knowledge and machinery that its audience actually could work with, or at least imagine working with—rational technological innovations that involved electric ovens, oscillating electric fans, autogenous soldering, or the use of a blowtorch.

Ciencia Popular—by contrast with a paper like *Crítica*, with its characteristic strategy of pursuing sensational stories—seems to have been a forum for technical realism, more geared to practical needs than to imagination plunged headlong into technological fantasy. But technological fantasy was not something it could entirely avoid when it was publishing articles about life on other planets (issue 9) and prospective trips to the moon (issues 22 and 23), or, in the guise of news items, stories that were on the same wavelength as science fiction. Nevertheless, Duclout, the editor of the magazine, like the thousands of readers it claimed (a claim backed up by the publication of photos showing trucks parked outside the press and piled high with copies of the magazine, or showing kiosks literally stuffed with *Ciencia Popular*), was disposed to technical realism, though *Ciencia Popular* certainly had Bovaryesque imagination as a major ingredient insofar as it featured experiments that could never have been performed in the environment of a home workshop.

Along these lines, two graven images that dominated the editorial page of *Ciencia Popular*—those of Jules Verne and Thomas Edison, a visionary and an inventor—reinforced this blend of realistic imagination and fictional oddity. Verne was the writer who had created a scientific landscape populated by successful explorers; Edison, the hero (typically American) of every technology-oriented publication. The triumvirate was completed, predictably enough, by Guglielmo Marconi, even though his likeness formed no part of this editorial-page pantheon. By contrast with these three, Einstein—the great, strictly contemporary figure, the sage theorizing about a physics whose hypotheses could never have crossed paths with an experiment conducted in a workshop—was criticized in issues 19 and 23 of *Ciencia Popular* with a virulence that was highly irregular for the magazine's otherwise measured, balanced tone of petit bourgeois didacticism. The ornamental frame around the table of contents repeated the motif of the future entwined with the present (or of imagination joined to practice): in the lower right-hand corner, a man fiddled with the levers of a gadget that was topped off by radio and television screens, but his stylized suit (recalling the vivid illustrations in *Amazing Stories*), like the rays striking the screens, divorced this motif from any kind of realistic interpretation without quite placing it in a completely fictional context.[20] At the top of the page was an airplane, more stylized and aerodynamic than existing models but an airplane nevertheless, even though it was flying across a background that included the vague outlines of planets. *Ciencia Popular* reached its apex with this illustration, which showed the magazine at its most audacious, and this audacity was probably necessary in order for a technology that was already on the verge of becoming ordinary to be included in a domain that was not about to part with its more imaginative features. The technology that the magazine offered its readers was almost always workable, if often baroquely complicated by bricolage and substitution of makeshift elements for serviceable ones; the accompanying bursts of creativity made up a minor corpus.

In this regard, the outlook at *Ciencia Popular* was closely related to the one shared by all the other radiotelephony magazines, which served as manuals and data banks for thousands of radio amateurs, but it was also closely related to the control over journalistic imagination that was wielded by a newspaper like *El Mundo*, less inclined than *Crítica* was to sensationalize information. What they all had in common was the burden of a body of practical knowledge that had to be transmitted to readers; the discursive genres of this knowledge were series of lessons, specialized publicity (in the form of minutely detailed advertisements, pedagogic in nature, from which one could learn, just as one could learn from catalogs), glossaries, and regularly published sections. Radio, photography, cinema, engines, electricity—whatever phenomenon such sections presented was already on the horizon of the ordinary, even though it retained its aura of novelty and modernity. The regular sections worked precisely like a set of hinges between inexplicable wonders and a technical manipulation that was not seeking to discover the laws of these wonders but to build and operate devices. Indeed, these sections dealt with utilitarian wonders that, already ceasing to be wonders, were taking their place within the predictable, normal framework of everyday life. But in the transition from inexplicable wonders to the utilitarian kind, newspapers and magazines were the discursive arena in which those who had radios and those who did not, those who possessed technical knowledge and those who merely enjoyed or marveled at its effects, could observe the dissemination of knowledge, facts, hypotheses, and projections of the future, which accompanied triumphs of technology as their cultural shadow.

Announcements of Things to Come

Crítica, with credulous voracity, positively ate up hypotheses about the future, and these cannot be understood in isolation from the paper's enthusiasm for technological and scientific modernization; on the contrary, novelties were links in a chain that became the very stuff of fiction—mythic aspects of scientific reasoning went hand in glove with processes of material innovation.[21] Anticipation of the future defined a register that was contemporary with the opening chapters of modern science fiction. Books and magazines devoted to predicting the future imposed a narrative or discursive structure on those fragments of the new that the papers were conveying in the dispersed (and often enigmatic) form of news items. The genre of the "fantastic voyage" transported the reader in time and constructed a panorama of what the future could be expected to bring, setting a scene of ordered novelty in place of a miscellaneous heap of previously unknown devices, substances, and procedures. In this way a spectacle was created, nourished by hypotheses that imposed order on the marvelous and indicated the direction in which it would advance; it replied to pessimistic views of technology—views supported by what the weaponry of World War I had brought about—with the optimism of a narrative in which science was rational and technology was working for positive changes in everyday life. If that was not the exclusive drift of such projections, it was still the predominant journalistic outlook, certainly unimpeded by Argentina's position of marginality and its comparatively slower pace of technological development.[22]

In North America, as we know, the relationship between scientific journalism and anticipation of the future began in the first decades of the twentieth century.[23] An equally systematic discursive structure cannot be claimed for Argentina, but from time to time *Crítica* did fashion information from the wire services into scenes of a future (sometimes as distant as two and a half centuries ahead) about which the reader was offered assurances that things truly would unfold in the hypothesized direction. Any difficulties that, as part of a generalized cultural processing, one might encounter in assimilating the new were countered with the idea that it was already possible to draw the future's major lines of perspective by tracing the hypotheses, based on real conditions, that were shifting the subjunctive mood to the future indicative.

Squibs and other short descriptive items functioned as a kind of down payment on the news; the Buenos Aires of 2177 would depend on the elaboration of versions of the city that were, as it happened, being supplied in the present by the newspaper itself. "In the Year 2177, Buenos Aires Will Be a City That Verne Could Not Have Imagined" is how *Crítica* headlined one major article on the future, with an illustration that occupied an entire page spread. It was the most comprehensive exercise in forecasting to be published by any of the newspapers and magazines of the period:

> Within 250 years, industrial control over the forces of nature will completely transform the way human beings live on the earth. This is the vision of our illustrator, who has predicted that this fantastic city, hundreds of stories high, will have hangars for enormous spaceships to guide million-candlepower reflectors into a sky that will never again be blackened by factory smokestacks. . . . Before a century has elapsed, discoveries in the field

of electricity will bring about the complete transformation of industrial development and living conditions. . . . In the city of the future there will be a fundamental change in how electrical power is obtained. Until now, methods of producing electricity have been mechanical, chemical, or hydraulic, but electrical current in the future will be drawn from the atmosphere. . . . All the heavy electrical charges in the atmosphere, circling the earth, will be condensed and distributed to industry. This distribution will not be carried out the way it is today, through endless networks of wires crisscrossing the cities; instead, highly powerful antennae will send energy out to the four corners of the earth. . . . These advances will bring about the complete disappearance of the automobile. . . . Tiny autogiro-style aeroplanes will allow city residents to go up to the 130th floor. . . . These electrical media, together with chemical advances using the explosive power of nuclear fission, will permit rapid transcontinental travel at speeds greater than 2,000 kilometers per hour. Cargo missiles will be shot up to great heights in space, and their descent paths will cover vast distances. . . . All of this will make it easy to cover areas the size of continents, with travel from America to Europe, and vice versa, accomplished in a matter of minutes. . . . Photographic transmission will be perfected to the point where the motion picture industry can be consolidated into one or two central stations that will transmit films, in the form of broadcasts, to all movie fans for the low price of a monthly subscription. . . . This, added to radiotelephony and the truly astonishing speed of train travel, will create new political conditions in the world. . . . States will be reduced to provinces administered by a universal council. The futility of monopolizing wealth, and a state of perfect communism, will establish an amount of individual labor that, being natural and comfortable, will neither demean people nor reduce them to servanthood in a regime where the great proletariat has to dive for the crumbs that fall from from the banquet table. . . . People will not live in the cities, which will be administrative centers and territorial granaries. Because fast train travel will erase distances, people will build their houses in the mountains, and in irrigated deserts transformed into gardens through the electrification of agriculture. . . . This, added to the simplification of industry, whose technical perfection will be almost absolute, and to advances in bacteriology and medicine, whose role will be prevention and immunization rather than healing, will make the dream of human happiness a reality.[24]

This great descriptive synthesis reached from the material world to the political sphere and the organization of society; the moral lesson of the future rode in on the notion of an egalitarian, happy world. Forster, Čapek, and Zamiatin saw nothing but nightmares in a technological rationality that would flatten out differences, stunt identity, and doom the individual; from the notably plebeian standpoint of *Crítica*, however, it looked as if the introduction of a system in which technological wonders would bring about progress and a social paradise did not mean hurtling into an abyss of authoritarian scientific rationality but rather throwing the doors open to a homogeneous, well-balanced society. A work of utopian fiction like *Ralph 124C 41+: A Romance of the Year 2660*, by the North American writer Hugo Gernsback, also belonged to the realm of popular journalism, where it served a purpose completely different from the one served by the technocratic nightmares of a Zamiatin or a Huxley.[25]

The point of this *Crítica* article was not to issue warnings about the dangers of technology divorced from political and moral values but to enact the technological perspective in a futuristic setting, which would be, in essence, a unified society that would exercise control over technological progress, knowing nothing of its negative consequences and achieving a political order of universal happiness—a future curiously identical to the one in the dystopian visions, but at odds with their belief that bringing technology to bear on life and social equality would bring about cultural impoverishment, moral bankruptcy, and psychological misery. What was emphasized, instead of warnings about the dangers of life's unlimited technologization, were the advantages of a social order in which work had been simplified, or perhaps abolished, by machines and by absolute mastery of nature. In fact, the *Crítica* article depicted a wage earners' utopia, one where they would have nothing to fear from a future in which the table at society's banquet, if holding less variety, would be completely accessible as well as technologically perfect.

The science fiction published in North American pulp magazines in the 1920s and 1930s was also more inclined to celebrate than distrust technological progress. These publications shared editorial and public space with hobbyist magazines devoted to electricity and radio, in whose blend of information, helpful tips, and didactic fiction readers found a pedagogy made to the measure of their intellectual training, to the world of work in which they lived, and to their aspirations for improving their daily lives. Such magazines presented technology in a way that made its understanding and handling seem possible even for those who had no systematic training of any kind, but who did have nimble hands, a feel for tools, and the capacity to learn the skills connected with practical material experience. The intellectual elites distrusted the leveling power of technological innovations (these were associated with mass democracies, or with socialist regimes that admired Fordism), and not every aesthetic vanguard shared the mechanistic symbolism of Russian cubism or Italian futurism; but what the proletariat saw in technology, apart from the mythic power of modern wonders, was its ability to improve everyday life.[26] Where technological predictions were concerned, newspapers and magazines aimed at this social sector were innocent of pessimistic hypotheses, as their audience also was; as a result, it is difficult to find any warnings in them about the uniformity and repressive effects of progress.

Crítica, *El Mundo*, and scores of magazines from the 1920s contained not even the shadow of a doubt about the news they were publishing. On the contrary, their attention was wholly trained on predicting still more pieces of what looked like an attractive future, one from which Argentina was thought not to be excluded. Technological and scientific anticipation announced future events with the certainty that they would be "universal," and that the city in 2177 would be the Buenos Airies of the *Crítica* article's headline.

Signals would be received from Mars and other planets. It would be possible to choose the sex of one's offspring and create life in laboratories. Gigantic airplanes would cross the Atlantic, and the floating cities built for them would also serve as models for space-exploration stations, which were to be constructed in conjunction with an expedition that in very short order would reach the moon. Classes would be offered via television, which would also broadcast surgical operations and transmit X rays, which doctors, at great distances from their patients, would use in making diagnoses. The police would use worldwide long-distance communication to exchange photographs of criminals and information about them. Transportation systems and urban lighting would be able to forgo the use of coal and petroleum by using clean atomic energy. Only synthetic food would be consumed, and ovens would be rendered obsolete. Robots would perform almost all the chores that were now the responsibility of laborers and housewives. Petroleum, wood, and other substances now obtained through archaic processes of extraction would come to be manufactured. Jet engines would replace combustion engines in almost all machinery and vehicles. No doubt wars, too, would be waged in settings where technology had multiplied the destructive power of weapons a thousandfold; but news of this kind had struck a deal with progress, and the terms of the deal left few possibilties for thinking about progress's moral limits or consequences.[27]

These predictions were based on the belief that the future would be better than the present, since scientific reasoning and its applications would fortify a number of desirable characteristics, such as speed, automation, egalitarianism, liberation from work, and world unity.[28] It did not occur to the mass-circulation newspapers and magazines (as it did to an important sector of the Western intellectual elite during the same period) that along the infinite road of technological progress one might also encounter unknown destructive power, not to mention possibilities for new forms of enslavement and alienation. Given the experience of the preceding decades—the legacy of which had been the paraphernalia of radio and cinema, two captivating modern miracles—neither the newspapers nor their public had any reason to suspect that material and technological progress, whose leveling force seemed to them something to be desired, might be more menace than blessing. Newspapers and magazines, reasoning by analogy, carried forward into the future the changes that technology had already introduced into everyday life: if this or that was possible, they asserted, then what seemed unbelievable in the present would also be possible; if this or that was good, then there was no reason to distrust future applications of science. This simple reasoning, based on judgments about experience, wiped memory clean of the machinery that technology had brought to World War I; thus new weaponry was presented more in light of deterrence than of aggression, and its technological up-to-dateness, expressed in terms of its speed, precision, and potency, was stressed more than its concretely destructive power. Intellectuals were the only ones warning about the dangers closing in on the future, and they wrote antiwar novels that found broader exposure in leftist magazines. By contrast, the journalism of technology saw the future as something more to be desired than feared as a threat, although the conflictive presence of both outlooks existed in a writer and journalist like Roberto Arlt.[29]

At any rate, optimism went well with the dynamic of papers that were chasing the news, and that were competing not just for news but also for its speedy delivery to the public. The Hoe rotary press at *Crítica* was the jet airplane of journalism; both belonged to a new era in which miracles were being incorporated into the implements of work and the objects of ordinary life, and especially to an era that was seeing a rapid and widespread increase in access to machinery, from radios to engines. This type of journalism approached history not just as depiction of the present but as prediction of its seamless continuity with the future. Moreover, the readers of this kind of journalism, to judge from the changes that the technological present had brought into their lives, had nothing to fear. Thus the newspapers, closing the circle, created and reinforced an interest also arising in other domains, which were linked more directly to experience, and which likewise had seen no dwindling of hope in the miracles promised not just by science but by a "beyond" of spiritual powers that found expression through parapsychologists, hypnotists, and clairvoyants.

The combination of all these features, each in its own time, exercised the attraction that belonged, if an oxymoron can be permitted, to the knowable mystery of modern wonders.

PART II

HISTORIES

CHAPTER FOUR

Inventors: Technology and Mythmaking

> Many a man roams the streets inflamed by a restless creativity, but most never see their dreams come true, because most of the time—stupefied by poverty, and lacking follow-up and perseverance or any real talent—they fail to ground their ideas in concrete reality.
>
> —*Crítica*, 1929

Technological Fantasies

In modernity, the present shows its strengths, not its temporal weakness, since the present holds the promise of the future. Amateur inventors, swept up in the whirlwind of this promise, believed that what they divined in their obsessional technological daydreams was the call to glory, some favorable turn of fortune, or a clear act of service to humanity. The relationship between the means they possessed and the ends they pursued pushed them into an unequal struggle for which they were only middlingly prepared; they often failed to distinguish between an idea that had plausible technological applications and some crackpot notion that never could have seen the light of day even as an experimental prototype. They read stories in the press about newfangled items apparently as outlandish as their own inventions, and when they turned up at newsrooms or the editorial offices of magazines, they were not always able to tell the difference between an accommodating reporter's irony and the sincere surprise he might express about the inventions for which they hoped to gain invaluable exposure, lacking economic resources and any real connections with formally trained technicians.

Only a small minority of these inventors had mastered any skills that went beyond the level of simply being handy, and those few constituted an elite who were not so desperately in need of newspaper publicity for their bright ideas, which were generally more in line with the actual possibilities for technological development. The others wavered between the desire for innovation and the limits of an expertise that had cut its teeth on handicrafts, practical mechanics, and a smattering of books.

The Arltian inventor—with a touch of the visionary and a touch of the swindler, a cross between the autodidact and the man on the make—constitutes a literary type whose explanation can be looked for in the places with which Roberto Arlt was intimately acquainted, as revealed by several of his journalistic essays, or "etchings."[1] Crazy inventors who presented themselves at editorial offices to promote their discoveries were dazzled by the image of the Wizard of Menlo Park,[2] who was, along with Guglielmo Marconi, a popular figure at the mass-circulation papers. The two of them presided over a pantheon of inventors, famous and unknown, that was flashed over the wires from Europe and North America to the newsrooms of Buenos Aires. Often news of this kind reported as a fait accompli some technological development that would not become possible for another twenty or thirty years. A certain fictional aspect enlarged the scope of such news, bringing together existing inventions, projections of the future, fantastic hypotheses, and the homegrown notions of skilled or brilliant amateur inventors.[3] A random sampling of such articles, but one that is not atypical in terms of their subject matter and the frequency with which they appeared, shows that in the first four months of 1930 *El Mundo* published international news about refinements of inventions, or new discoveries, in areas as varied and fascinating as medicine, photography, radio, cinematography, weaponry, dirigibles and airplanes, television, long-distance photographic transmission, lighting, optical signals for maritime communication, new chemical elements, refrigeration, wireless telephony and telegraphy, jet-propelled autos, and new fuels and energy sources.

In Buenos Aires during the 1920s, hundreds of individuals were caught up in this dizzy whirl. Perhaps one of them had designed a radio the size of a ring; another, a new prototype for a battery-operated wristwatch, or an automatic file that could simplify office work, or a 3-centimeter galena radio built into a deluxe case and another radio small enough to fit inside a walnut, or a metal that had the advantage of replacing the galena used in transmitters, or an "airmobile" whose distribution was being blocked by problems with the patent, or a device for making an engine's movements reversible, or an instrument that improved the possibilities for long-distance reception, or a machine that automatically made change.[4]

The inventor is a social type throughout the West, and especially in the United States,[5] but to encounter him in Argentina allows us to imagine how the cultural thrust of modernization, and especially of its myths, gained ground in a country that, if relatively marginal (and, of course, intellectually removed from the trend for innovation that was sweeping the north), was not unaware that the nineteenth-century cultural elite's ideology of Spiritism was being countered by another ideology: Americanism, a better fit with what seemed desirable in the plebeian world but also in the world of the capitalist movers and shakers, and in the world of those who were founding new states, such as Soviet Russia.

At the same time, the inventor was seeking several things that had no direct connection with scientific activity; by contrast with the researcher who remained unknown in his lifetime, the inventor wanted recognition, fame, and wealth. These desires went hand in hand with technological invention, which had links not just with the practical world but with financial success and upward social mobility. The typology of the inventor included, very fundamentally in Argentina, the amateur of humble origins. The Argentine inventors were not invention-lab employees like those in North America, where veritable innovation factories had sprung up; they were enthusiasts of the new, competing in extremely precarious technological conditions and in an environment swarming with autodidacts.

Nevertheless, they did sense that they were in a position to compete, even though almost none of them had achieved any success as an amateur inventor. After the newspapers reported the experiment performed by Marconi in which wireless signals had been transmitted a distance of 17,000 kilometers, one Doctor Amallo, general director of the Mail and Telegraph Bureau of Argentina, addressed himself to the newspapers to make it known that, as *El Mundo* quotes him,

> this splendid experiment, conducted with such obvious success, nevertheless came as no surprise to those in the bureau who are quietly devoted to studying and perfecting this kind of activity.
>
> Indeed, as the enclosed document shows, in the month of July 1929 Mr. Rodolfo Paolucci, chief of the Telegraph Office, submitted to the division of technical experiments of the Army General Office of Troops and Communications Services a machine of his own invention that is intended to produce, through the emission of Hertzian waves, various types of mechanical effects, such as the lighting of electric lamps, the incandescence and fusion of compatible wires, or the operation of devices adapted for all kinds of auditory or sound effects.[6]

Paolucci's experimental trials may have had some connection with Marconi's experiment, but other creative locals (who introduced themselves in newsrooms with business cards on which "Inventor" appeared under their names) built cumbersome prototypes that had been brought into being without so much as a nod to the notion that their full realization might entail technical difficulties. In the town of San Martín, two friends drunk on reading Jules Verne and H. G. Wells went to the authorities to propose the construction of a machine that would produce "heat in the air, up to a million units of pressure, in order to conduct scientific research on the space between the Earth and the moon." An automobile mechanic waxed enthusiastic before a none too skeptical reporter as he described the method he had just dreamed up for achieving vertical flight without a propeller. Another told of having invented a submarine, a prototype of which he had built from balsa wood, whose distinguishing feature was that it carried a satellite boat for lighting the wicks of the attack cannons.[7] A few who received mention in the papers seemed more like international adventurers, such as the young Neapolitan academic who claimed the ability to produce electricity for industrial and household use through the chemical reaction unleashed by the contact of two "special pastes."

And there was also the draftsman Parpagnoli, who spent years working on a device for creating a gradual change of speed in electrical and combustion engines. A former colleague of his, writing about him in *Crítica*, said, "Unlike others, with their bohemianism, this painstaking Italian draftsman devotes himself to the noble task of making a contribution to the progress of mechanical science." According to the article, Parpagnoli had accomplished this goal and now needed only the patent and financial backing so he could move to the United States, where his invention was sure to make him a millionaire.[8]

Parpagnoli's problem—to develop an invention that, for the moment, existed only on paper—was one for which other solutions were being attempted by the skilled do-it-yourselfers of Buenos Aires, thanks to the beaches littered with mechanical trash, which one of Arlt's essays placed in the 10000 block of Calle Rivadavia. These beaches were haunted not only by competing amateur inventors but also by the all but useless remains of engines (along with scalpers of spare parts); adventurers predisposed to the planning of sea voyages, for the sake of whose theoretical achievement they devoted themselves to transforming canoes into ships; farmers looking for small electrical generators, which might release them from the prices imposed by the local factory; professional mechanics; and neighborhood repairmen. Here is Arlt on the inventors:

> There goes a gentleman who has invented an appliance. He doesn't have the money to get it manufactured. That doesn't matter to him. He'll use automobile parts to build one that looks something like it. Here comes a citizen convinced that he has invented perpetual motion, who now lives in terror that "they" will "steal" his "secret," and who, to keep "them" from "stealing" it—how sweet! how naive!—goes out and buys himself a 150-horsepower engine and shuts himself in his room with it and gazes at it and coddles it and thinks his discovery will be a sensation, once he gets it up and running. . . . And here is a penniless inventor who heads for this necropolis as a form of educational entertainment whenever the going gets tough, coming here with a look of resignation to ponder the junk and reflect that he, too,

> could finish his invention if he had the money. . . . These are the men we see here, like exiles from some mechanical paradise, long-suffering beings who, if they could, would tuck an engine under one arm, carry it home, lovingly put it to bed, and tear it to pieces in order to "improve" it.[9]

Invention was a cultural topic, and the inventor was a type to whom Arlt's writing would return, and not just in the ironic style of this essay. These "exiles from some mechanical paradise" did know about the existence of *this* paradise, and that was all that mattered for the time being; their love of technology and their dreams of success knew no bounds in the presence of this disorderly, disparate collection of almost totally useless parts. Obtaining a patent was known to be very difficult,[10] just as it was known that the building of the invention (a phase for which professional inventors in North America used actual laboratories of experts, who had access not only to parts but also to banks of chemical, mechanical, and metallurgical data) would make a diabolical labyrinth out of the act of obtaining legal recognition for it. And yet the urge toward technology is what fired this group that lingered over engine parts and, even more important, over the papers, which continuously published news about inventions, raising the audience's expectations and encouraging audience response.

The inventor was respected as someone who had made a contribution to a common culture, sometimes an implausible contribution, or one marked by the autodidact's unwieldy complexity. Right alongside the frauds, who were unrelentingly investigated by newspapers acting in the public interest,[11] were draftsmen like Parpagnoli, or simply builders of attractive scale models whose clever workings got their inventors into the papers for a day—characters like the old man concerned about the astronomical education of children, for whom he had built a model of the solar system, which, together with the study of planetary movements, would allow the perfection of the Divine Plan to be verified.[12]

Clearly, there were no precise categories for inventions, discoveries, the dissemination of science and technology, technological experiments, or research. In the news of Argentina, and in stories from the wire services, everything fell into place along a single continuum—the new world of the twentieth century, a world that many articles made it their business to declare closer at hand than it actually was. It can be assumed that when newspapers examined this continuum, but failed to establish major differences among the various types of news, the impact on readers was likewise undifferentiated; thus *Crítica* and *El Mundo* were approached by builders of prototypes, by visionaries who had reinvented the wheel or introduced complications into communication systems about which they understood nothing at all, and by hopeless cases who were trying to compete with foreign patents by showing reporters the evidence of Argentine registration. At one extreme on this news continuum were the Arltian inventors and—products of a kindred discourse—some of the characters in Arlt's novels.

In spite of it all, the didactics of modern journalism did attempt to impose some kind of order on this technological domain. That was the purpose of the special sections, which combined simple explanations with tips for building devices, or with highly technical disquisitions on valves and circuits. But this order was constantly disrupted by articles that could not do without sensationalism, or by the celebration of some imminent advance, or by contamination with the dubious neighboring zones of insanity and delusion. The same amalgam can be assumed on the part of these articles' consumers, especially in social strata where, mythically speaking, such articles' imaginative impact was fostered by the absence of formal education and institutional support. In the first chapter of *El juguete rabioso,*[13] the boys who rob the library make up a fellowship not just of thieves but of inventors. They are prepared to sell a portion of their plunder, but they show that they are fully aware of what the scientific manuals are worth.

Necessary Information

These amateurs, with the exception of the intellectual elite who gathered at the Círculo Argentino de Inventores, had not received their technical education at universities. It is easy to picture them learning on the job in industrial or quasi-industrial businesses, in machine shops, in the studios of the earliest ham radio operators, and at the workbenches of handymen who were already repairing radios and other electrical devices, or one can imagine them simply studying machinery through the store windows of Buenos Aires, like one of Arlt's characters.[14] They haunted the neighborhood libraries, where technical books were in great demand,[15] as well as working-class training institutes and the expositions devoted to radio, cinema, and engineering. They signed up for correspondence courses, which targeted not just their desire for knowledge but their fantasies of economic triumph and upward social mobility; that, at least, is how the ads of the 1920s and 1930s made their pitch:

> Do not confuse the situation of the simple employee, forever doing the same work, or that of the laborer, indentured to the workshop, with the situation of the man who has been trained for a professional career, the man whose knowledge adds up, day by day, to success. The most capable employee is the Office Manager. The one who has been acquiring knowledge ever since his workshop days becomes the Technician, and the progress of the whole factory takes its direction from him.[16]

The offerings of these curricula (and their proliferation indicates the commercial success of their sponsors, if not of their enrollees) included "career" training as well as degrees that evoked and sometimes literally mimicked those of the university (which, the reader was assured, every professor responsible for teaching the correspondence courses had attended; no one, the reader was assured, could embark on an apprenticeship without knowing how to read and write). The most frequent offerings were technical courses in radio, mechanics, and electronics, which covered antennae, circuits, valves, capacitors, transformers, resistors, loudspeakers, chassis, combustion engines, boilers, turbines, hydraulic and steam engines, automobiles, generators, and AC/DC motors, as appropriate. Training as a lathe operator was offered at one point on the labor continuum; at another, training as a "mechanical engineer," an "electrical engineer," or even a "railroad engineer." A typical photo in one ad showed a smiling youth in a dark twill suit and irreproachable shoes; with a handkerchief in his front pocket, a diploma under his arm, and his right hand resting on the lever of an electric motor, he is highlighted against a background of enlarged cogwheels:

> The bright outlook for the field of automotive mechanics means that there will be constant demand for technical experts. Automobiles, trucks, and tractors continue to catch on with every passing day, and this development naturally increases the need for repair technicians with solid professional ability.[17]

At any rate, articles and illustrations like these can be seen as having given momentum, from the angle of theory, to those who had already had contact with modern machinery in home workshops or hobbyists' sheds. Such prior contact was assumed (and sometimes its limitations were pointed out) in course advertisements, which informed prospective students about the differences they should have been able to perceive between the simple assembly of a device (a skill apparently assumed to be widespread) and the theoretical understanding of basic radiotelephony as it applied to solutions for practical problems that a novice who only worked empirically was not even remotely equipped to figure out. Thus there were abundant guarantees that "general notions" of physics, mechanics, chemistry, electricity, and electromagnetism would be imparted, and these were presented as a body of knowledge that was superior to other schools' mere "know-how," which one could acquire in its entirety without ever needing to enroll in any of those institutions.

This type of know-how, both among amateurs and among technicians who had been getting independent on-the-job training in the 1920s, is precisely what combined with materials of varying provenance and difficulty, from the college texts and popular-science books available in public libraries to advertisements for correspondence courses, to produce an accumulation of different grades of knowledge, a circulating amalgam that constituted the ideological and imaginative substrate of the world of professional technicians but also and especially of those who were not professionals. Shoptalk, magazines and their letters from readers, and continuous newspaper coverage of these issues marked off a new and not completely institutionalized zone of knowledge, one in which even those who lacked formal educational credentials could move about freely. And they were the very people who became consumers of this kind of news, as well as of these courses and what they had to offer.

Among all those who made their way through this web of discourses and technical myths, which blended beautifully with the desire for upward social mobility and cultural legitimacy, some, even many, did become inventors. For them there were books about invention, such as the *Manual del inventor*, by T. A. Le Breton, and several editions of the engineer Elisa Bachofen's highly successful *Guía del inventor*.[18] Le Breton's *Manual* was openly committed to the idea that the great modern inventions had come about through the efforts of beginners who had no advanced training in science and often no practical experience in their areas of invention.[19] What was more, many successful inventions had been accidental. Finally, it was claimed, inventive ability was innate, and even if it did depend on certain kinds of technical knowledge, it did not need them as much as the systematized professions did (that is, professions that were preceded by a formal apprenticeship).

In short, anyone could be an inventor, but the stroke of luck and genius behind every completed invention was sustained by the efficacy of certain very simple rules, which the *Manual* set out as a means of preventing failure where a resounding success might be achieved instead. In the first place, it was necessary to avoid "imperfect execution," caused by the inventor's anxiety to get his invention out into the world. The *Manual* cautiously advised waiting as long as necessary rather than exhibiting a defective prototype. The other golden rule said that the foundation of the most profitable inventions was observation of the most ordinary actions; whence the corrugated hairpin, which had brought its inventor half a million francs and had come about through his daily observation of his wife as she did her hair, twisting the pins so they would not come loose. According to the *Manual*, the same thing had occurred with the mousetrap, the safety pin, plush towels, juicers made of glass, and barbed wire. The lesson of these inventions was a hymn of praise to the household wizards who had been able to look at exactly what needed to be seen, and to see it in terms of innovation, even though their innovations involved mere details. On this point, the household handyman had to make the effort to link his technical means (generally limited) with clear needs for improvement in everyday life or at work, and with the prospects for his invention's commercial distribution. The *Manual* imparted the fact, no less valuable for being simple, that every good invention contained the possibility of financial gain; the inventor's fantasies would have to be ruled by the limits of industrial or technical demand if the inventor hoped to profit from his discovery rather than be counted among the unlucky, who generally died poor and unknown, and whose inventions made fortunes for others who did understand how to align them with the technology and economy of the times.

The *Manual* criticized the impatience of amateur inventors, especially their ambition for every single one of their inventions to be a hit with industry (it was noted that Edison had obtained patents for fourteen hundred inventions, only four hundred of which found practical application). Above all, the *Manual* drilled its readers on three questions that should structure their efforts: How could the cost of manual labor be reduced? How could work time be reduced? How could better use be made of raw materials or by-products?[20] Every worker or employee could ask himself these simple questions; employers, moreover, were clearly prepared to reward inventions in the workplace, as proved by the hundreds of examples cited in the *Manual*. Apart from this fact, though, which tended in any case to highlight the benevolent good faith of industrial and commercial employers, it was important for amateur inventors to realize that they had no need of any knowledge or understanding that lay beyond their reach; inventions were right there in everyday life, the products of shrewd observation applied to ordinary problems, which, just because they *were* ordinary, seemed to have less to do with the thrilling world of cutting-edge innovation. Invention was possible on the scale of the workshop, the office, and the home (and here the *Manual* highlighted the hundreds of patents that had been obtained in other countries by women determined to improve the performance of household tasks).

This aspect of invention may not have been so appealing to inventors of sensational objects, who were in the habit of paying visits to newsrooms, but it did reassure handymen who were insecure about the kinds of knowledge that might be indispensable to invention: innate ability was to be perfected less through ultra-sophisticated types of scientific knowledge than through technical knowledge gained outside formal institutions, and, above all, it was to rely on the power of observation, which would discover, in the folds of the ordinary, the gaps that invention could fill. On this point, the *Manual* gave particularly reassuring examples of what a necessary invention was, even though such an invention might have been so ordinary as to seem deceptively simple, and even though the answers to unsolved problems might have been difficult to find in a context of homespun technology.[21] Nevertheless, these propositions in the *Manual* served two purposes: they defined the minimum requirements for invention, in terms that were different from those used for systematized scientific learning; and they outlined problems that were immediately understandable (if perhaps unsolvable), using simple explanations that reflected everyday experience. Thus invention remained imaginatively inscribed within a familiar framework that could be encompassed by the more or less basic skills of the amateur engineer, worker, or handyman who had the gift of inventive talent. As the *Manual* informed its readers, a number of purely domestic annoyances and preoccupations had led to lucrative inventions. And if financial success could come to those who followed particular steps, then once an idea had "appeared," the reduced scale and the homegrown nature of the process were reinforced by the advice to work on what was ordinary. The fact that things did not actually work this way, anywhere at all, was a limitation that neither the *Manual* nor hundreds of Buenos Aires inventors were prepared to acknowledge.

And this was all the more true of those whose imaginations took them beyond, and set them against, these simple recommendations. The inventors who hung around newsrooms, those Arltian characters in quest of perpetual motion or floral galvanization, took to heart only one of the lessons being offered: that *anyone* could be an inventor, and the less formal learning he had, the freer his skills and flights of imagination (or delusion) could be.

It is difficult to know who these inventors were, apart from the members of the Círculo Argentino, who can almost be followed on a case-by-case basis, for there were only about 150 of them, with perhaps a few more outside that elite group (although their inventions met the same fate as those of the local working-class inventors—before Biro, with his ballpoint pen, not a single successful patent had originated in Argentina).[22] The fact that the number of patents in Argentina had increased is not necessarily an indication of the number or the caliber of Argentine inventors, much less of their fate. In any event, the record-breaking year of 1923, when 2,893 patents and local renewals of patents for international inventions were granted, saw three times the number that had been recorded in 1911, and 1923 continued to be the high-water mark for another ten years.[23] The patents encompassed businesses as well as independent inventors, but the nationalities of the applicants convey some idea of how many of the inventors were local and thus, no doubt, independent: 1,269 of these inventors were Argentine.[24] Names are repeated in the files of the Office of Trademarks and Patents, and so the number of inventions is somewhat larger than the number of inventors, but only occasionally did an inventor patent more than one invention in a single year.

And yet obtaining a patent was a complicated, expensive activity. Not every inventor reached that dreamed-of moment of finding himself deposited on the threshold of wealth or prestige (and a dream it certainly was, since it was unusual for an invention to find its way into industry unless a business obtained its own patent). Again, the newsroom habitués, who found it all but impossible to build a prototype, threw theirs together from a hodgepodge of discards and spare parts. Patenting an invention entailed a descriptive memorandum, a technical drawing, and the fee that had to be paid to the Office of Trademarks and Patents. The obstacles piled up when a descriptive memo was too vague for the conventional bureaucratic language, and inventors scurried back and forth between their workshops and the patent office, finding themselves only halfway down a road on which they had started out believing that the most complicated part—the invention itself, that difficult and sublime achievement—was already behind them.

For these inventors, intimidated by the obstacles to obtaining a patent, there was the friendly world of magazines like *Ciencia Popular*, which took for granted the existence of ideas and projects that could be turned into useful applications on a scale that was limited and, intellectually speaking, no doubt unambitious as well.[25] The magazine sought practical suggestions for small improvements, ideas that clearly were not in competition with great inventions but might prove to be the seeds of a fortune, as the *Manual del inventor* had already declared, or that could, at any rate, change ordinary life for the better. The "Practical Ideas from Readers" page included a range of items: a device for slanting a mirror so it could be adjusted to the angle required by its user's height; a hanger for trousers; and—a typical feat of do-it-yourself ingenuity—an oscillating fan constructed from a pair of wooden propellers rotating on large armatures.[26]

The letters column gives the impression that there were also plenty of contributions involving more complex inventions; the magazine's replies to its correspondents were peppered with advice concerning the need to perfect and test the ideas being proposed, which ranged from technical improvements to unadulterated tinkering. The instructions offered by *Ciencia Popular* apparently ruled the imaginations of the magazine's readers, although it is impossible to know what was actually being sent to its editorial offices. The marvelous inventors featured in *Crítica* were not the ones in *Ciencia Popular*, who constituted a world of dreamers prepared to work within somewhat narrower limits in their quest for unspectacular improvements. Sometimes the magazine's advice went beyond technical issues, venturing into the psychological and epistemological conditions of invention. But readers were haunted by one question: Was it possible to be an inventor if one did not have a solid scientific education? The answer was generally equivocal: if there was no wish to suggest that it was possible to be an inventor when specific kinds of knowledge were totally lacking, neither was there any wish to exclude from this universe of small improvements (or—who knew?—great strokes of luck) the hundreds of readers who had confessed their limitations in the very act of posing this question. An example:

> Our sincere congratulations on your desire to better yourself and be of use to humanity! Unfortunately, however—as you yourself have so aptly put it—not much can be done without a foundation of fairly solid training. All the same, willpower can carry you to success. It was not because of his education and training that Stephenson, the creator of the steam engine, became successful. Or, for another example of willpower, consider Gillette, a man of the people and the inventor of the famous razor, who waited more than eighteen years for an investor to show interest in his invention; in time, he became famous.[27]

The idea that other inventors, inventors who were now famous, had also suffered humiliation and slights at the hands of capitalists who were blind to progress, or of reactionary scientists whose distrust was matched only by their utter lack of originality, was one motif of the "lives" that appeared at fairly regular intervals in these specialized magazines and newspaper sections. The model of success was certainly Edison, who had been lucky enough to gain early recognition and to win his independence by building his own invention lab. The focus was on Edison, since he had validated an invention myth that revolved around financial success and social renown; but inventors like Stephenson, Morse,[28] and all those others who had been written off as crazy, or scorned by official science and by industrialists as shortsighted as they were greedy, were not forgotten, and their stories just might foreshadow the reader's (after all, there was no telling when one might cease to be misunderstood and take one's place in the charmed circle of the public spotlight). This discourse was comforting to obscure inventors, and it held out the promise of a brighter future than the decent technical job that one might obtain through a working-class training institute or a correspondence course. In fact, future rewards and public recognition in newspapers and magazines could be put on the back burner as long as one could see that other inventors, those famous unfortunates, had gained fame only after having endured circumstances as humiliating as they were depressing. The life stories of inventors, rather than emphasizing how difficult it was to be successful, proved that success was possible. Through these life stories, hundreds of amateur inventors, and thousands of readers with an interest in technology, discovered the possibility—no less imaginary for all its resemblance to their fantasies—of seeing themselves in the self-made men of other countries.

Technology had no privileged place among the intellectual elite; it was a relatively democratic field in which know-how played an important role, compensating for lack of education and creating a connection, actual as well as imaginary, between professional engineers and the Sunday handymen toiling away in their home workshops. Magazines and newspapers reinforced these ideas with the simple language they used in presenting their material, and with the exemplary life stories they offered their readers. The history of technology in Argentina was not a long one; and, in a way, all those who took on technical problems were taking off from the same starting point. Technology allowed facets of the imagination, which knew no class differences, to be put to work, and it favored the members of the working class, placing them at an advantage because of their manual skills and their mastery as artisans or tradesmen.

An Elite of Inventors

The Círculo Argentino de Inventores brought together individuals of another kind. Founded on December 28, 1922, by Colonel Adrián Ruiz Moreno, an engineer and an excitable, egocentric, nationalistic character, the Círculo assembled an elite of university-educated inventors who kept up a lively dialogue with politicians and political organizations (especially those of the far right) as well as with manufacturing and business corporations, fashionable clubs, mainstream journalism, and a variety of governmental agencies.[29] About three weeks before the organization's first meeting, Ruiz Moreno published an article in *La Nación* in which he explained a few of the reasons why the Círculo had been founded. On the one hand, Argentine capitalists were unresponsive to the forces of progress, which could be the driving force behind a mutually profitable relationship between themselves and inventors; moreover, the government, which was just as apathetic as the capitalists, knew nothing about the efforts of those who were putting their intelligence and their lives at the service of humanity, and it was humanity, after all, that would "truly enjoy the benefits." On the other hand, the absence of laws for the protection of inventors was reinforcing both this indifference among the rich and the lack of curiosity on the part of government officials. Ruiz Moreno's protest was based on the perceived need for invention to be considered a patriotic activity, since it was creating possibilities for industrialization in a country that should not, in his opinion, be content to remain an agricultural power, as had also been demonstrated by the needs that appeared during World War I and afterward; and invention was creating these possibilities at the very moment when, again in his opinion, population growth was making it necessary for such possibilities to be opened up to those who were, as he put it, "landing on our shores."[30] Although the Círculo readily defined itself as a society that was technical in nature, the other adjectives customarily used to describe it were "cultural," "patriotic," "scientific," and "altruistic." In his remarks at the Círculo's inaugural meeting, Ruiz Moreno repeated almost verbatim the arguments he had made in *La Nación*, underlining both his summons to the capitalists, who needed to understand that it was morally superior and financially more productive to fund an invention than to buy a showplace in the countryside, and his call to the public authorities in defense of *national* inventions, which implied some kind of strategy when it came to the renewal of foreign patents in Argentina.

The founding members numbered thirty-three, emblematically enough.[31] Ruiz Moreno was elected president, of course. Among the thirty-three there were five engineers, and three of them served on the first steering committee.[32] Several engineers, as well as military technicians from both the army and the navy, were mentioned in the first publications and meetings of the Círculo,[33] and almost immediately two women engineers joined—the Bachofen sisters, one of whom, Elisa, would serve as secretary and editorial director.

Right away there was a meeting of the bylaws committee, also presided over by Ruiz Moreno (the majority of its members were academics), and the requirements for membership in the Círculo were established. A member could be "any creator of useful, patented inventions, or anyone who has been publicly recognized as an inventor for a period of at least one year, or who may be accepted by the technical subcommittee of the Círculo." Aspirants born outside Argentina were required to have been in the country for six years, to have "worked during that time or produced something useful," and to "indicate settled residency or real property, demonstrating the same with the appropriate credentials and good conduct." To these basic qualifications,[34] which outlined the universe of "serious inventors," a provision was added by which the thirty-three founders would maintain permanent active membership (and therefore eligibility for the steering committee), whereas a new member would lose active membership after going five years without producing an invention. In a subsequent article of the same bylaw, the phrase "regular exercise of the profession" again stressed the requirement for continuous rather than occasional invention.[35] Monthly dues of 3 pesos were established, 10 percent of which was allocated to the creation of a relief fund "intended to be of assistance to members and their families" in the event of sickness or death, or when "through unusual circumstances an inventor [should fall] into indigence . . . or when he [should] suffer . . . an accident while testing his invention."[36] Consideration was also given to issuing credits for project development (from the beginning, stubbornly and with scant success, the Círculo lobbied banking institutions for a system of preferential credits) using money that, it was thought, would accumulate in a savings account to which would be added donations, bequests, and members' transfers of royalties on inventions. Among the objectives stated in article 1 of the bylaws were the creation of a cooperative for medical and legal services, the building of a social center and a mausoleum, and the development of a library. These were the classic functions of any association. The unprecedented goals of this one were to promote its members' inventions, facilitate the procurement of patents at home and abroad, manage the distribution and sale of inventions,[37] promote the passage of a law to protect inventors, and ensure compliance with patent laws and the honoring abroad of local patents.

This radically new corporation had a corporate-defense program that presupposed a membership that was capable of surmounting the obstacles of the invention-patenting process and coming up with the $82 tax levied on a five-year patent. But it was probably not this payment (which could also be made in annual installments) that fundamentally separated the members of the Círculo from inventors who were "not professionals." Instead it was the bureaucratic and technical gauntlet that had to be run before the patent was attained, and which itself represented a corpus of skills that were elucidated from time to time in the Círculo's magazine.[38] This regulatory maze outstripped the proficiency of amateur inventors and pragmatically oriented handymen. In reality, the Círculo straddled the fence that had been erected by the Office of Trademarks and Patents, pointing out the official institution's arbitrariness and inefficiencies but also accepting the repertoire of conceptual skills that it demanded of inventors. Those members of the Círculo who were engineers, led by Ruiz Moreno and the Bachofen sisters, were in a position to help inventors who were comparatively well prepared and better educated, but the Círculo did not in any way present itself as a service agency for inventors who were operating far outside the complexities of a technology that had to be transferred onto paper in abstract form, not merely subjected to hands-on physical testing by way of machinery or prototypes.

Thus the Círculo represented an elite of inventors, but one that continued to grow—by 1923, there were one hundred forty-three active members in addition to the thirty-three founders.[39] And activity in the treasury during the first eight years (1922–1930) indicated a sixfold increase in new memberships. In the same period, the value of inventoried goods went from $160 to $10,938.[40]

Other traits of the multifaceted, egocentric Colonel Ruiz Moreno may have contributed to the profile of the Círculo's members, who tended to be socially respectable and relatively cultured. Ruiz Moreno himself chose the Círculo's motto ("A Cultural, Patriotic, Scientific, Altruistic Society") as well as the members' distinctive medal: on the front lounged the figure of the Nation, seated on a rock, sword aloft and illuminated by the rays of a rising sun; the obverse bore the legend "Honor and Country." The meaning of this image, more ideological than technological (the medal made no reference to such objects as the pen, the hammer, and gears, or to other, equally predictable iconography), became clear when the Círculo's magazine published news of Ruiz Moreno's attendance at conferences sponsored by the Liga Patriótica Argentina. Although the Círculo's political independence was repeatedly emphasized, no one in a position to write for the magazine seems to have been troubled by the activities the president undertook in representing the Círculo, which also garnered praise from the Liga Patriótica's committee on culture and education.[41] There is nothing else to suggest any ideological affiliations on the part of the Círculo, which indeed seems not to have had any. But Ruiz Moreno's style, in addition to his ideology, may have lent pomp and formality to the everyday life of the organization, a certain stiffness to the observations that the magazine directed at the Círculo's members, and an uncommonly stately quality to its gatherings for amateur inventors, such as they can be imagined from the letters columns of the specialized publications.

But Colonel Ruiz Moreno's egomania left its mark in the form of two other organizational features, which were certainly quaint (or, to say the least, highly unusual with respect to republican customs, whether elite or plebeian): in Ruiz Moreno's honor, Founder's Day was established, and on that day he was specially regaled (although the other founding members were permitted to bask in the glow of this homage); and the Círculo's magazine took its improbable name—*ICARM*—from the acronym formed by the initials of Ruiz Moreno's name preceded by the initials that stood for his title and his military rank: *ICARM*, for Ingeniero Coronel Adrián Ruiz Moreno.[42]

The Círculo's activities were impressive in their persistence as well as in their systematic nature. Annual contests, together with an exposition of Argentine inventions (held at the Círculo's headquarters, or at the fairgrounds of the Sociedad Rural, where Ruiz Moreno received President Marcelo Torcuato de Alvear and members of his cabinet), got the organization into the daily papers and, on a regular basis, into specialized publications like *Radio Cultura, Radio Revista*, and *Ciencia Popular*. But the themes of the Círculo's contests, along with medals offered by national ministries and provincial governments, and juries composed of professionals and official personages, were what eased the same insecurities among amateurs that were addressed by the newspapers and the most popular of the specialized magazines. These themes could be said to be "serious"; they were organized into rigorous categories, and what the list of prizewinning inventions evoked was not the world of puttering handymen but that of a more systematic technology. In general, the winning inventions were useful and plausible innovations, and they followed the counsel so often spelled out by the Círculo's magazine in its "Problems Solved and Unsolved" section, a genuine guide that steered the reader clear of dead ends and amateurish fantasies alike.[43] The inventions promoted and exhibited by the Círculo had less to do with strokes of luck than with the politics (which never proved successful) of soliciting the interest of Argentine industrialists, who failed to respond as the Círculo had hoped they would to a strategy that might guarantee "economic and industrial independence based on a national enterprise of invention."[44]

In this respect, the elite members of the Círculo were different from the amateur inventors, even though neither group achieved resounding success. The inventions that won prizes from the Círculo were feasible, were responses to "realistic" themes, and were presented in all the detail that was called for by the rules of technical drawing and prototype construction. A far cry from the often unworkable brainstorms of handymen—whose only technical education had taken place in factories and workshops, when it hadn't come from magazines—the inventions that emerged from the Círculo sought clear differentiation from the world of amateurs for whom invention was not only a hobby but also and especially a new kind of poetic imagining. Here, then, were two socially defined worlds, which certainly intersected—at the expositions, if not at the fringes, where the Círculo might lend a hand to this or that do-it-yourselfer who needed to systematize his invention—but which were distinguished by their methods of invention as well as by the desires that technology unleashed in them.

CHAPTER FIVE

Radio, Cinema, and Television: Long-Distance Communication

> The paved alleyway ran between tall fences and tightly shuttered houses. A few damp brick courtyards gave off bursts of a moist, oregano-scented freshness, and a distant radio was playing somnolent music that was interrupted from time to time by a rooster's high-pitched song.
>
> —Roberto Arlt, *El amor brujo*

From Ham Radio to Broadcasting

A photograph of Ezequiel Paz, editor of *La Prensa* and president of the Radio Club Argentino, occupies the cover of the January 1923 inaugural issue of *Radio Cultura*, which doubled as the magazine and broadcast schedule of the eponymous station that had been established on May 29, 1922, to fulfill the "dream of bringing the loveliest expressions of art and thought to all the residents of the Republic." The covers of succeeding issues featured a gallery of individuals involved in radio. There was Juan Quevedo, who chatted with Paz by radio and was prepared to broadcast, from his own home, the phoned-in requests of those who had already purchased radios and were now looking for something to listen to. There was the architect Ovidio Carpinacci, as ready as Quevedo was to play requests: "We feel like dancing, Señor Carpinacci. Play us a foxtrot, and some tango music after that, and then a waltz!" There was the engineer Guillermo Guntsche, author of a popular manual on wireless telephony for amateurs, who had been a contributor to *Radio Cultura* from the first issue, and who did justice to the way it presented him (as a "wise and intelligent young man"). There was Luis Orlandini, a navy captain who, after being stationed in Austria, brought back with him the novelty of radio and became the driving force behind the founding of the Radio Club Argentino. There was the engineer Teodoro Bellocq, "a rich man who studied and educated himself as if he were an intelligent poor man," whose fifth-floor residence in San Isidro[1] housed the nation's finest radiotelephony lab, where Bellocq had been picking up North American broadcasts since 1914. There was Ignacio Gómez, the youthful *Radio Revista* editor, who in the joint opinion of *Radio Cultura*'s editors was an industrious role model for ill-mannered "rich kids." There was Rafael Mastropaolo, who served as the Radio Club's first secretary but was obliged to relinquish that office in order to become assistant manager of an "important company," no doubt the same one that made him director of a radio exposition at Luna Park.[2] There was Luis J. Romero, a partner with César Guerrico, Miguel Mujica, and Enrique T. Susini in what were already legendary musical broadcasts from the Teatro Coliseo. And there was Jorge Duclout, engineer and, with Guillermo Guntsche, coauthor of a book on wireless telephony, a topic to which Duclout would return several years later, this time as editor of *Ciencia Popular*.[3] All of them had been organizers and were now members of the Radio Club Argentino, established on October 21, 1921, when (as they wrote three years later) "radio was still in a state of gestation" and no one could have predicted that "it would develop so suddenly, and that things would happen so fast."[4]

These pioneers belonged to the middle class of Buenos Aires or, in several cases, to the city's upper middle class (one illustrious scion or another might rub elbows with the now settled, second-generation sons of Argentine immigrants for whom things had gone well), but the movement they led became a mass phenomenon, almost without anyone's taking note of this social transmutation. When

Radio Cultura was boasting twenty thousand copies per week, that figure may have been inflated to attract advertisers, but it still spoke of an audience that, six months after the first issue, could already be asked to pay 20 centavos a week for a magazine that until then had been free.

In September 1923, the magazine's cover photos of the top men in broadcasting gave way to a domestic scene: the household of Guillermo Cartagena, 1,300 kilometers from Buenos Aires—in Mendoza, perhaps, or Patagonia, or Salta. The photo shows Cartagena, a "radiotelephony fanatic," next to his wife and a little boy of six or seven, who is sitting on the floor on a small round rag rug. Father and mother, seated in wicker armchairs, one on each side of the child, have their eyes fixed on an indeterminate (and, by definition, unattainable) point. The child, with equal concentration, is looking at the camera. The three of them are listening. Señora Cartagena is using one set of headphones while Cartagena himself shares another set with their son. On the table is a radio that Cartagena must have put together, for it has no cabinet and the tubes are visible, as are wires leading to the capacitors and pickup coils, the sphere of a galena radio dial, and various needles. Radio Cultura offered two daily programs, one broadcast by the Palermo station (this one, specializing in dance music, was the more popular of the two) and the other by Sudamérica (it featured concerts of music from the most familiar classical repertoire as interpreted by the faculty and students of the Buenos Aires conservatories, by the Trío Radio Cultural, and by the Orquesta de Radio Sudamérica). But transmitters, receivers, and accessories were also advertised under the Radio Sudamérica banner, and one of the official distributors was Radio Cultura, with offices at 600 Callao.[5] Señor Cartagena, who clearly had built his own radio, belonged to a circle in every sense of the term: in the early years, radio coupled technical occupations (and preoccupations) with the new enjoyment and excitement of long-distance communication; magazines and broadcasters formed a network through which one could find supplies, advice, practical information, and an audience for one's experiences and problems; and the listeners were not just technicians but also potential broadcasters. They all faced the same problems, exchanged foolproof formulas, and gave as much advice as they took.

In 1923, radiomania was fueled by a variety of publications, by the aforementioned exposition at Luna Park, and by the same informal network of technical enthusiasts. Radio was a dream come true, in which the technical met the marvelous precisely at the point where an artfully placed antenna made it possible to listen to Pracánico's[6] tangos or Señorita Noacco's violin rendition of a Grieg berceuse. The circuit fed on solutions to technical problems and pleasure in radio reception, and this pleasure was technical in nature, for it was to this that Señor Cartagena turned in his leisure time, reading manuals, writing to the letters columns of various magazines, sending away to warehouses for pamphlets and catalogs, mail-ordering parts for his transmitter, and no doubt repairing the odd coil or despairing over a burned-out tube. In advertisements,[7] the prices for receivers and parts ranged from 300 to 1,500 pesos, but it didn't take much to improve a homemade radio, and, as the magazines counseled, it was best not to spend money on wooden or ebonite cabinets, for the amateur builder could expect to be carrying out almost continuous structural modifications on his receiver.

"The Ficaseosté Family Under Attack" is the caption for the cartoon gracing the cover of the September 1923 issue of *Radio Revista*.[8] The family has nine members, every one of whom is totally given over to lunatic radiotelephonic tinkering. Señor Ficaseosté, using the back of a wicker chair, is attempting to mark and punch out the holes into which the base of the radio will be placed. He has already perforated the board with a manual drill, and the cartoon shows him plunging the drill bit into his dog's hindquarters while staring at the disaster that his wife and an aunt have caused by overturning the pot of shellac they have been using to secure the wire wrapped around a pickup coil. Señora Ficaseosté and her sister, the guilty parties in this minor misfortune, are distracted by the armchair onto which the bodies of the two little Ficaseosté girls have just fallen. The girls themselves, completely absorbed in wrapping the wire around the coil, were performing their task with such diligence that the wire snapped under the strain, and both girls were almost thrown to the floor, but one of them was not, precisely because she used her mother to break her fall, thus causing the pot of shellac to tip over. Two uncles, somewhat younger than the head of the family, are the authors of the disasters in the background. One has just smashed his hand with the hammer he was using to nail a wooden cabinet together. The other, responsible for carrying the poles of the antenna out to the terrace, but distracted by the prevailing uproar, has hooked the parrot's perch with the end of one pole and shattered the sideboard mirror with the other. In the foreground, the family cat swats at a vacuum tube while the more tranquil grandfather and his little grandson read *Radio Revista*. The Ficaseosté family certainly must have attended the Luna Park expo; in any case, another cartoon in the same issue shows a girl and her little brother among the curious throng gathered around a dubious expert who is explaining the difference between galena and a vacuum tube to an evidently middle-class audience whose members are examining an impressive bank of generators and looking for the wires and plates that power the exhibit's broadcasting station. There are also dozens of booths that feature the same advertisements as those published in radio magazines and heard on the radio.[9] No doubt there were also amateur handymen at the expo, such as Domingo Castiñeiras, who built a miniature galena receiver ten times smaller than a telephone headset, or the do-it-yourselfers that the notes on the exposition alluded to with the headline "A Series of Unusual Things," which appeared under drawings of three objects—a pumpkin, a shoe, and a cage—that had been made into radio receivers.

But it was not just magazines specializing in this area that reported these inroads by creative handymen. During the week of the Luna Park radio exposition, *Crítica*, which was to broadcast an informational bulletin over the expo's closed circuit, announced the most sensational news of all: the world championship boxing match between Luis Firpo and Jack Dempsey would be heard over a wireless broadcast; Western Electric had built special loudspeakers, and the racetrack annex would be made available, with "comfortable seats and boxes where families [could] easily find places."

But there was more: the Radio Club Argentino was sponsoring an amateur competition into which an "enormous number of radios" had already been entered.[10] The local creative types were typically handymen who had a miniaturist, kitsch aesthetic:

> The miniature receiver measures 3 by 4 centimeters and is mounted on a base of Brazilian pine and imitation ebony. Its slim, muslin-lined box makes it look more like a deluxe item than a versatile electromagnetic-wave receiver.[11]

> Vicente Rodríguez Viergol Figueroa, a young man carrying two small galena receivers, introduced himself to us this morning. One of his devices, constructed with no little skill, is housed in a small box like the ones commonly used for potassium chloride tablets. The other . . . is a precious ring of silver. The receiver is mounted in the central portion of the ring and consists of two telephone jacks, two tiny grounding connections, and an antenna. Señor Rodríguez Viergol Figueroa has sharpened his talent to the point of creating a piece of jewelry, and one doesn't know which to admire more: the arrangement of the compact attachments, or the tiny device itself. The little box and the ring have been used to listen to every broadcast in the metropolitan area. Both receivers will be on display at the International Radio Communications Exposition that starts the day after tomorrow at Luna Park.[12]

> When Little Mario Aragone snacks on nuts, he doesn't just toss the shells. He uses them to build radiotelephony receivers as efficient and perfect as the originals that his copies reproduce. He has mounted a small, powerful receiver inside the lower half of a nutshell and fashioned a fine cabinet from the upper half. The nut/receiver consists of a galena tuner, a coil, and the four terminals that go with it, two for the telephone and two for the antenna and the grounding connection. This tiny, original little radio has been used to listen to every station in the metropolitan area, especially those of the ham radio operators.[13]

Radio was the wonder of wonders, and there were many enthusiasts in those early years whose relationship with it was one of enjoyment fostered by proficiency. Indeed, according to a 1926 editorial in *Radio Revista*, "There are not just seven wonders of the world, nor are we surprised anymore by the automobile, the dirigible, lethal weapons, the railroads, or transatlantic liners," as if listening to the radio were not just a passive activity, and as if the new technology that was so much in evidence—in those years, it was already possible to tune a radio—had taken fantasies from the literature of prognostication and made them come true, revealing them to be not utopian technologies but a potential reality that within a few decades would be part of ordinary life.[14] "Compared with the automobile," the editorial writer went on, "radio has possibilities that even today have yet to be explored."

What was the source of this unexplored, supposedly infinite potential? It was the very nature of the technical medium, its literally fantastic character, which gave material form to hypotheses that until then were believed to be fictional, "marvelous." The "miracle" of radio was talked about no end, and this metaphor is the sign not so much of excited exaggeration as of the practical realization of a myth—communication with the unseen, the vanquishing of the senses' corporeal, material limitations—that the collective imagination associated both with the paranormal and with the technoscientific. There was a link in the popular imagination between immateriality and the power of radio. Radio waves were invisible to broadcasters and listeners alike. One could enjoy the transformation of electrical waves into sound waves, and vice versa, without having the slightest understanding of the principles involved. Moreover, a sharp distinction could be drawn between the scientific knowledge that made radio possible and the technical knowledge that went into handling the components of a homemade radio, or between scientific knowledge and one's feel for positioning a needle or an antenna, or between scientific knowledge and the luck that governed this intersection of nature and technology by determining whether there would be excellent reception or none at all. It was not the specialized publications or special newspaper sections that sought to make this sharp distinction between science and technology; they were doing their best to explain the physical principles behind radio broadcasting. The distinction was made by radio enthusiasts, as all the letters columns of these publications make abundantly clear. Technical difficulties inspired virtually all the questions submitted to such columns, and the replies revolved around technical solutions. Entire issues were devoted to "beginners,"[15] and a certain number of the readers of these publications, specialized to the point of being obsessively monothematic, were considered to need only as much instruction as might improve their ability to fiddle with a radio's dials, since they themselves disregarded the idea of building their own radios, although the desire to do just that seems to have been remarkably common, if the amount of advertising devoted to the assembly of homemade radio receivers is any indication. And, as shown by the price ranges of the components that were for sale, radio enthusiasts of different social backgrounds and economic means could choose from a wide variety of products, encompassing both the engineer Teodoro Bellocq's laboratory in the canyons of San Isidro and the radios assembled on kitchen tables for a total cost (excluding headphones) of 35 pesos, a sum that might be spent over a period of two weeks as each part was mastered with the aid of instructions and drawings in the manuals that were also copiously advertised, and that were as inexpensive as the era's mass-market paperbacks.[16] There were classified ads, too, which filled a regular section in *Ciencia Popular* and offered used radios, parts, and tubes as well as services like rewrapping a coil. The know-how absorbed from sales catalogs put the finishing touch on the education provided by magazines and by the special sections that certain newspapers, such as *Crítica,* had been publishing since the early 1920s. The magazines, of course, given their pioneer ethic, advocated building one's own radio, and yet their pages also ran ads for ready-made radios that cost the same as the homemade ones, but without offering the financial advantage of an installment plan of ten or more payments.

An ethic of technical pioneering is also evident in the types of contests that the same magazines sponsored for their readers, who thereby tended to become the magazines' collaborators, reinforcing the notion of a community of interests in which there was no sharp differentiation between the two wings of technical journalism (in fact, one of the magazines had the motto "Written By and For Amateurs"). The lists of prizewinners reveal the strongly practical outlook of the judges and contestants, and there is no reason to expect anything else.[17] There was even speculation that this primacy of technology over science was a distinguishing characteristic of the type of knowledge, and skill, that was at issue:

> Successful transatlantic broadcasts by amateurs have offered clear proof of the great efficiency of shortwaves, even though this level of efficiency was not, until now, regarded as compatible with shortwave-broadcasting theories and formulations. Once again, the amateurs have proved not only their technical mettle but also the fact that calculations and formulas have only relative value in the new science of radio, and that practice and experimentation, above all else, are what will lead to new discoveries. Some time ago, governments relegated the radio amateur to wavelengths of less than 200 meters, judging them to be of little use or profit in radio communication. But the determined experimenter, whom stringent laws compelled not to exceed this limit, managed to devote all his intelligence to improving the receiving devices first, and then the transmitters, and in this way, through trial and error and considerable effort, shortwave radio devices have undergone great improvement.[18]

Which encapsulates the conflict about expertise. Apart from the monopoly that governments exercised over radio waves and their allocation—a prerogative intolerable to the anarchistic spontaneity of the radio pioneers, and one they continually harped on—there was the little drama of the conflict between scientists, that is, between academics, who were not self-taught and who belonged to the intellectual elite, and technical amateurs, who basically were self-taught, learned from manuals rather than institutions and from peers rather than teachers, and formed horizontal relationships with other amateurs rather than hierarchical relationships, even when a hierarchy existed, as defined by one's seniority as a ham radio operator or by one's evident dedication to the public duties that allowed amateur radio to advance (as in the case of the more or less elegant elite that ran the Radio Club, even though that dedication in no way exempted the members of this elite from the criticism customarily leveled at them by the magazines).[19] It was a classic conflict—science or technology? Or, in the terms of its manifestation in the social imaginary, who were the real drivers of progress, and where were the obstacles?

The answers fell along fairly basic populist lines. On the one hand, governments were butting in where they had no business, behaving unfairly toward the independent initiatives of the powerless "little guys," and damaging their interests by selling their abilities short. On the other hand, however, governments had also seen—no small irony—that the work ethic of the amateurs, even those who were poorest and least educated, could make a mockery of governmental objectives. The allocation of shortwaves had placed worldwide radiotelephonic connections squarely into the amateurs' hands. And not only that, the amateurs had achieved this outcome precisely because they had proved not only that science was wrong but that it was wrong in its attitude toward the repetitious yet innovative power of experimentation. This was knowledge versus know-how, the world of intellectual culture versus the small neighborhood workshops (and the occasional workshop in a San Isidro villa or a military barracks).

At this stage, the Radio Club elite was apparently not a social world that was completely closed to the working-class amateur. The club's sponsored activities were not exclusive, and even though the logo illustrating the organization's official bulletin depicted a private club of sorts—elegant in the English manner, with men in formal dress, as if filing out of a theater—it is possible to imagine that these distinguished amateurs crossed paths (if not wires) on the air with the twenty thousand readers of the specialized publications, along with their friends and families. At this stage of radio broadcasting and reception, the airwaves were a democratizing force. Moreover, the radio expositions with their thousands of visitors, held year after year at Luna Park, at the Sociedad Rural, or in various central locations,[20] created a network of contacts in which the expertise of local working-class amateurs (the prizewinners in almost all the contests, as revealed by their addresses and surnames)[21] held its own with that of the elite and the educated bourgeoisie in the shared hobby of amateur radio. The influence of working-class know-how was both qualitative, as shown by the contests, and quantitative, as can be understood from the number of people who visited the expos and constituted the reading audience for pamphlets, catalogs, and periodicals. Radio's impact, a technological and modern one, extended far beyond an elite that can be shown not to have been radio's center of gravity.

The endless dispensing of practical advice also speaks of a world of amateurs whose economic resources were not unlimited, and who had to manage with recycled materials and do-it-yourself shrewdness. One obsessive question turned up time after time in letters from readers: How might tubes, plugs, coils, and headphones be scavenged? (The reply invariably pointed out how unlikely such finds would be.) Articles about converting one type of gadget into a new and improved or more powerful one; countless explanations of how to build parts that could have been obtained ready-made from the commercial outlets that were forever advertising them; shared news about conserving materials—all of this invites reflection, along two lines.

On the one hand, there was the artisan/amateur/do-it-yourself ethic of recycling, of using rubbish and discarded parts, broken and retooled items, objects redirected from their original functions, of impossible repairs that defied practical intelligence and manual skill. Success, according to this ethic, necessarily meant making the most of limited means through practical determinations (there were technological limits as well as technologies that were unavailable to home workshops) and symbolic economies. On the other hand, a small workshop did not contain "a little of everything"; there was what there was: used parts, tools not always wholly suited to the task at hand, conditions of permanent but not desperate need, of insecurity without wretched poverty. This is what can be inferred from the solutions that amateurs shared with one another as they overcame their common obstacles. The economic resources available to average amateurs—if indeed they can be distinguished from the two or three dozen elite ones whose names repeatedly come up in connection with the steering committees of the Radio Club Argentino—were limited by what appeared to be an acceptable amount to invest in a hobby. It cost 7 or 8 pesos to replace a headphone, 8 or 9 to get one's hands on a tube, 1 or 2 for the capacitors. Sums like these were not out of reach, but the constant proliferation of equipment was still not easy for middle- and working-class handymen who had given in to their passion for radio, and so any money the do-it-yourselfers and expert amateur mechanics managed to save could only reinforce the ethic of conservation. The ideology of these handymen and creative technicians was not based on unlimited expenditures but on endless recycling and retooling. Theirs was an art of fine-tuning. Their troubles began after the radio was built—the antenna could not be positioned, or the galena's conductivity was weak, or the dial was unresponsive, or a tube had burned out, or the valves had failed—and all these problems showed up in the practical advice and pleas for help that were found in the letters columns of newspapers and magazines. The world of home workshops (and even of tradesmen's workshops in working-class neighborhoods) was the world of unintended consequences and on-the-spot solutions. This type of homegrown innovation was as fascinating to its practitioners as it was to those who came up with ethical reasons for promoting it.

The public face of radiomania was quite sane. In the radio magazines, this cultural trend did not show what must have been its other face: fantasies about lucky inventors, or about sensations created by major discoveries, or about riches and fame raining down from the heavens on clever locals. No doubt such themes were peripheral to the specialized magazines, but they can be discerned in the calls paid to newspapers' editorial offices on behalf of what were generally far-fetched inventions. Nevertheless, the obsessive, patient radio ham was not such a frequent visitor to that open ward of a waiting room occupied by manufacturers of miniaturized gadgets, melters of metals, builders of ovens and boilers, and inventors of chemical processes. Next to them, the ham radio operator—painstaking, accustomed to spending his nights listening long and hard for any sound that might literally and metaphorically come from the beyond, by way of the ether—was an expert of sound mind, the quintessential pedagogical model for the middle and working classes.

The world of amateur radio and basic radio receivers got off to an energetic and technologically promising start, but within a few years it had split into at least two main segments. On one side were the ham radio operators, those methodical nocturnal seekers of messages flung into the ether by their fellow hams. Here, the technical possibilities were similar at the broadcasting and receiving ends, involving processes of back-and-forth communication whose end points could be thought of as symmetrical. Fantastic tales about this kind of symmetry were taken on by publications like the one put out by the Radio Club. For example, one Señor Landó and a certain Señor Chescotta, two radio hams from Argentina, had arranged one afternoon for both of them to tune in separately to the shortwave broadcast of a third ham in Rio de Janeiro. Apparently the two Argentines were interfering with each other's reception, and so one of them, with no particular thought in mind, began tapping his radio panel with a pencil, and these taps were heard by his partner, who replied in Morse code. Astonished by this sudden conversion of their radio receivers into transmitters, they both approached their panels and held a conversation. The Radio Club ran the story,[22] hailing this newest occasion on which technical prowess had managed to demonstrate facts that "science" had yet to discover. In any case, and regardless of the story's authenticity, experiments like this one (and like the ones that must have taken place every day among radio amateurs who owned transmitters and receivers) were proof of the broadcasting circuit's bipolar nature. Technical experimentation was not just a developmental phase; it was also a circuit of symmetrical articulation. Skilled amateurs, assemblers and dismantlers of household gadgets, and readers of the old and newer technical magazines[23] would continue to belong to this segment.

On the other side was the group composed of mere listeners, who had bought their ready-made radios at Casa América or Max Glucksmann, and for whom radio was a pastime, not a hobby. Everything on this side was about the audience for radio broadcasting as a business, which in the early 1920s may also have included amateur-run enterprises—such as Radio Cultura in the beginning, which used the services of amateur performers, musicians, and announcers from the conservatories—but which very soon became professional commercial radio stations, taking perhaps no more than five years to do so. In Buenos Aires in 1928, more than fourteen enterprises with authorized commercial frequencies had studios downtown that were broadcasting every day between noon and midnight. The most powerful of these was the municipal station known as Radio LOV.[24] For this new, nontechnical kind of radio listening, which would become the foundation of a dynamic cultural industry in the 1930s and 1940s, the newspapers offered advice about choosing programs, in addition to explications ("The ABC's of Radio," for example, in *El Mundo*) whose purpose was not to turn people into radio-building craftsmen but into a listening audience aware that its listening was the product of a technology whose basic principles, whether or not one aspired to master them, could be grasped by anyone.

This was the point at which the technology became opaque to its users, when the wonder that was radio, which for skilled amateurs had resided where technical aptitude met symbolic enjoyment, was fundamentally converted into a fantasyland divorced from technical concerns. The radio fanatics (or, as the Radio Club preferred to call them, the radiophiles) soldiered on in their workshops, but now alongside thousands of living-room and kitchen-table radios that marked the beginning of a new era in which radio technology was in the hands of the stations. Hence, several years later, the debate among the elite of radiotelephony pioneers about the Radio Club's proper functions; this divergence of opinion showed not only how many more radio amateurs there were but also how many more commercial radio stations.[25] The wonders of technology had coalesced.

The Silver Screen and Its Trappings

> We will then have reached the third stage of development, combining sight and sound in perfect fusion, possibly in color. And we will then have talking pictures, in natural color, and in this way we will achieve the closest possible representation of real life.
>
> —*Ciencia Popular*, 1929

In business, technology, and advertising, sound films and movies in color were major motifs in the second half of the 1920s. When North American talkies began to arrive in Argentina,[26] everyone, from film professionals and local movie studios to the Buenos Aires city council, felt called upon to offer an opinion on the subject. With some trepidation, the city council presented two plans to ban sound films, one because of how they would affect the local musicians' union and the other because of professed objections to public broadcasts in a foreign language. This knee-jerk reaction, a predictable and inevitable intersection of cultural nationalism and corporate interests intertwined with local fortunes, gave rise to a more generalized rejection, one to which arguably the most important popular technological magazine of the era devoted an editorial:

> This magazine has often published news about advances in sound films, since this kind of news is of interest to the general public. We will not enter into a discussion here of whether audiences in Buenos Aires prefer talking or silent pictures, although good talkies have certainly enjoyed a level of success unprecedented in the history of film, especially when we consider that they have been films in foreign languages. Why take a stand, then, against the normal course of progress?[27]

The question was primarily rhetorical. Toward the end of the silent-film era, sound was being talked up first of all from a narrative and aesthetic standpoint[28] but also from the point of view of technological development, as was well known in Buenos Aires, where, ever since the middle of 1920, papers and technology magazines had been reporting that sound films were on the way. By 1924, Phonofilm technology had earned quite a lengthy article in *Crítica*[29]—an interview, long on technical analysis, with Lee De Forest, known among wireless hobbyists for his invention of the three-electrode audion tube, which he had tested at the Eiffel Tower in 1908, thereby making radiotelephony possible (at least according to the breathless interviewer). As for what became known as De Forest Phonofilm, the method consisted in synchronizing sound directly onto film, a technical solution that had been the focus of experiments, and that would be the solution finally chosen by Hollywood, at least for a time. Three years after its adoption by the industry, De Forest Phonofilm arrived in Argentina, courtesy of the Argentine-American Film Corporation. In 1924, local film studios began screen tests with this innovative technology, which pointed up, as it had also done in Hollywood, the difference between a star's photogenic and phonogenic qualities. Sofía Bozán[30] submitted to what was apparently a successful screen test, and the magazine that published this news accompanied it with a photograph of a film set decorated in rural/gaucho style.

That magazine was *Ciencia Popular*,[31] which in 1929 produced a twenty-minute 16-millimeter documentary on the Puerto Nuevo factory of the Compañía Hispano Americana de Electricidad; the caption for one of the article's photograms makes an allusion to the avant-garde culture of the period ("A detail from the boiler room, which looks more like a vision from *Metropolis*"). The idea was to organize a "popular-science cinematheque," and the topic of the next film was to be a visit to the radio stations of Buenos Aires. A late–1920s Argentine documentary film industry would have been fascinating, but the magazine did not return to this project in the years that followed.

Nevertheless, *Ciencia Popular* did provide very thorough coverage of the movies, mixing technological wonders with Hollywood mythology. Given the great variety of journalistic genres, it was not uncommon for different types of news and information to be scattered across the same page. An informative article about sound might appear next to a complicated, step-by-step description of the successive stages of film manufacture, from the planting of the cotton whose fibers provided the base to the preparation of the emulsion and the loading of the light-sensitive film into the camera, and the page might conclude with the translation of a Hollywood gossip column in the form of a section titled "News in Brief." In this hodgepodge of information, the news about the movies always foretold progress without end. Sound, which until recently had been experimental, was now presented as a technology that would successfully arrange and present various auditory levels as "soundscapes," more striking and authentic than real life itself. Articles were often illustrated with images of the new armored cameras, and the captions persistently emphasized the armor's function of insulating microphones from the sounds produced by the intermittent mechanism that advanced the film. Directors, reduced to silence on the set, used gestures to indicate what they wanted in a performance instead of voicing their wishes as they had done in the era of silent films. Finally, there was the ultramagnification of the image, the blowup: the three cameras used to great effect by Abel Gance in his film *Napoléon* ("triple, synchronized projection"), experiments with larger-than-life sound films in 3-D, the use of negatives 63 to 70 millimeters wide, and 56-millimeter Magnafilm, which employed a special lens for printing and looked like 70-millimeter film when projected, all of which would soon make the 35-millimeter projected image look small.[32]

The heterogeneity of this journalistic material reveals how different film's status was from that of radio. Indeed, the articles' unorthodox layout—which turned the topic of "the movies" into a ragbag of technological curiosities, explications of incomprehensible processes, and trivial gossip proffered by foreign wire services that transmitted the glow of the star system—indicates that, by contrast with radios and other objects, which a moderately knowledgeable amateur could handle, film, even in contexts that were clearly more technical, initiated a highly mythologizing relationship of consumption, disrupted only by incursions into homemade production, and especially projection, of movies on the part of a few rich amateurs, most of whom quickly left the scene as soon as a local film industry began to coalesce.

But amateur filmmaking (certainly a much more expensive hobby than radio) still had enough visibility in 1930 for the Asociación Cinematográfica Argentina to sponsor a contest.[33] The previous year, readers of *El Mundo* had learned about the potential, at the fingertips of anyone who owned a typical home-movie projector, to view their movies in color. All they needed was a tricolor filter, with one blue strip, a strip of red, and another of green, turned to a new purpose. If the use of such a filter hardly seems practical, the reflections that accompanied the article had the prophetic tone that marked the style of the period, which was all about disseminating technological innovations; movies in color were "the future," of course, and the future was "starting to push the present aside."[34] A turn of the page revealed the newspaper's inspiration for this article—the Société Française Cinéchromatique had just been granted renewal rights in Argentina on a foreign patent for an item with these very features.

But amateur filmmaking—unlike radio, which bridged social classes—remained confined to "comfortable mansions," as *El Mundo* put it, of which there were nevertheless enough for Kodak to offer rentals from its library of movies, comic shorts, and animated cartoons, and enough for even the mass-circulation newspapers to keep coming back to the topic of amateur filmmaking every time they mentioned an exposition or some new innovation that a well-known company had imported. Whereas radio had created not only an audience but also, and even before that, a widespread network of amateur technicians, film created an industry almost from the start, along with a world of spectators whose only relationship with cinematic technology was by way of the imagination.

Therefore, articles like the one just described cannot be understood exclusively in terms of their marketing function; and, likewise, advertisements constructed a panorama of images and desires without bringing any do-it-yourself technicians out of the woodwork. The reasons were quite simple, and they were based on the fact that filmmaking was a far more expensive hobby than radio, as well as on the fact that it was virtually impossible to build one's own equipment.[35]

On the rare occasions when readers might find, say, an article containing a few suggestions for building a movie camera (which might double as a projector if its intermittent mechanism were removed and its reel system loosened), these suggestions, unlike the detailed instructions for building a radio, were unbelievably vague. A stark contrast with the kind of do-it-yourself information that focused on radio, they were instead an indication of how important movies had become in the imagination of an audience that now consumed them in dozens of commercial theaters. No one had systematically thought these suggestions through so that someone else could put them into practice. In reality, this kind of article was more likely to be an explanation of how film was imprinted so as to be projected at a speed of sixteen frames per second, or of how the intermittent mechanism pulled the light-sensitive film along, or of the reasons why the plate supporting the film had to be thoroughly polished, or of why it was necessary to avoid camera vibrations, and so on. And, in reality, nothing could be built with the aid of instructions that belonged to the comparatively imprecise world of creative technology. The point, then, was not to provide assembly instructions; using instructions like these to assemble anything at all would have posed an almost insurmountable challenge even to local handymen who had already demonstrated their prowess by building box cameras for still photography. The point was to explain how things worked, as if explanations like these represented the steps to be followed in a work of artisanship that could actually be carried out. Such explanations, more than supporting any type of real activity in home workshops, fed the imagination and promoted a certain level of technical knowledge about cinema. At the same time, however, alongside the classical journalism that had inspired the Hollywood star system and a local pantheon, there was a great deal of material that did discuss the technical dimensions of film, splicing and blending journalistic genres and slipping back and forth between them. This kind of material was of no practical use to the plebeian amateurs in their little neighborhood workshops, or even to the middle-class amateurs devoted to radio and photography, but its proliferation proves that it was necessary. The new audience, with which journalism maintained a dialogue of expectations created and confirmed, was drawn both to the mythological dimension of cinema and to its material and technological dimension (to the latter, not because of any practical potential it might have had, but as a source of information). An article titled "A Visit to the Casa Valle Laboratories"—with the film historian Chas de Cruz in the role of tour guide, walking and talking with the lighting designer Alberto Etchebehere—was illustrated with photographs whose documentary look and feel belonged to the genre of popular science.[36] The darkrooms with their vertical marble tanks, the area for washing and turning the negatives, the huge drums for drying them, the titlers and Moviolas, the printers, and, finally, the image that carried the mythical dimension—a projection room for directors, actors, and technical personnel, complete with art deco–paneled walls and wicker armchairs like the ones on Mar del Plata boulevards or in California houses—this was, in short, the silver screen, the domain of contemporary myth but also of imaginary visits to studios and trips to film sets and tours of laboratories with their Moviolas and, above all, their optical-effects tables. This industry had no room for the kind of artisanship that had opened up the worlds of radio and, earlier, photography. The movies created an audience that *watched*, and precisely what the movies constructed was a gaze and a way of seeing, detached from the skills that had connected radio amateurs to practice. Because the myth of cinema was not an exclusively aesthetic one, explanations like the ones offered in such articles fostered the technical imagination even as they exposed the limitations of the new cinematic audience's knowledge and aptitudes.

Divining the Future

If cinema's constant refrain was "The future is here now," television was a hypothetical technology that remained out of reach, no matter how many times the topic came up. Why, in the second half of the 1920s, did newspapers and magazines place so much emphasis on wireless long-distance transmission of images? To judge from the news, television was already an established fact. "Television Is a Marvelous Reality," read a *Crítica* headline in 1928,[37] and two years earlier the paper had published this triumphant squib:

> E. F. Anderson, a consultant for the Radio Corporation of America, has announced his invention of a device for the wireless transmission of cinematic images.
>
> According to Anderson, an engineer, it will be possible to broadcast moving pictures across the Atlantic after his invention has undergone further improvement.
>
> Anderson's invention, called a "television set," operates twenty times faster than a telephoto device, needing only one-sixteenth of a second to display an image, whereas a telephoto device requires twenty minutes.
>
> Pictures are projected onto the screen by a light ray vibrating at a rate of 300,000 times per second and reflected by twenty-four rotating mirrors.[38]

In the same year as its headline proclaiming the "reality" of television, *Crítica* ran a front-page story on this "most recent and curious electrical wonder," explaining in minute detail the process that had made "an old dream come true." From his laboratory outside London, the Scottish engineer John Logie Baird had sent shortwave signals to a station in Hartsdale, New York, where they were received in the presence of a reporter from the Associated Press. And these signals were not only seen but also heard over an ordinary radio: "They came across in the form of a musical hum with an irregular beat, and its tonal variations followed the waves and flashes of light that played over the faces of the people whose images were being transmitted. We might call this an experience of 'hearing vision.'"[39]

The news of this technological wonder came wrapped in metaphor. It was not just that the images had been seen; their very nature was sonorous, and thus they were turned into something as familiar as radio, which supplied the analytical basis on which the existence of television could be understood and, in particular, made plausible. At any rate, newspapers as well as their readers were prepared to believe in this new invention, and especially in how quickly it would spread, as had been the case with radio. Everyone wanted to believe that television was about to leave the laboratory, and, even more, that the idea of a homemade television set was practically a guaranteed reality, since all the parts, with the exception of a "special tube," were available from radio stores, although the retail cost of what was presented as the first homemade television set—that is, one that could be used outside an experimental lab—would be around $400.[40]

Television's potential—to be a home theater of images received at a distance—was greater than radio's. But two years after this burst of sensational optimism in *Crítica*, not only had television failed to arrive, there was also discussion in the United States about how long it would take to develop it beyond the experimental stage and make it into what cinema, radio, and the phonograph had already become. These time estimates were optimistic, in any case; no one thought it would take longer than five years.[41] It is surprising (and suggestive of readers' credulity and desire) that the newspapers were so emphatic about the speedy advent of the home television set, which was expected to complete the circuit opened by radio, and to marry radio with film, in a historical arc encompassing the two technological wonders of the first third of the twentieth century. Journalists and their public were reasoning by analogy in this instance: the introduction of radio, and the control over radio technology that was exercised by ham operators and radio stations, encouraged the belief that long-distance image transmission would recapitulate the evolution of radio. The reality of the 1950s would cruelly demonstrate how wrong it had been to assume this kind of symmetry. Meanwhile, among those who knew little or nothing about the technology involved, it seemed axiomatic that television equaled cinema plus radio, although there was absolutely no factual basis for such an assumption.

But even in specialized magazines and manuals where information about the technology did exist, television was spoken of as the "undisputed reality" to which *Crítica* had referred, and it was also called "telecinema."[42] As a combination of moving pictures and long-distance transmission, it brought together two technological developments that had proved to be not just completely feasible but relatively inexpensive (as radio was) and quick to catch on (witness the dozens of Buenos Aires movie theaters). There was nothing—apart, perhaps, from the specific character of television itself—to indicate that this history could not be repeated.

Local experiments began right away, and they may also have been driven by the idea that a world war over patents had recently broken out, and that the battle could be joined by Argentine engineers.[43] And it was not just engineers but local merchants, too, who were fantasizing about this newest of big businesses. In the wake of several television experiments presented at the International Radio Communications Exposition of 1929, the materials and implements for "having your own television at home" were widely advertised in Buenos Aires stores. One of the specialized magazines set aside its customary technological optimism to denounce this near fraud :

> We recently saw a story in the paper about a North American firm advertising a home television set for less than 100 Argentine

> pesos. We have conducted television experiments in our laboratory (and we continue to be the only ones to have presented experiments of this kind to the public), and so we believe that this also gives us the right to state our opinion. During the last radiocommunications exposition, at the Teatro Opera, about 60,000 people filed past our television set, and of these 60,000, only about 1,000 had any real understanding of television. . . . To get better results from our own set, we had to spend large sums on expensive equipment in order to solve problems that, mathematically speaking, can be shown not even to have solutions at the present time. Today, those who advertise a television set for 100 pesos are preying on the unsuspecting, hoping to sell them a worthless object.[44]

But people wanted to believe that television had already arrived, and the same magazine that denounced the opportunism of merchants, not to mention the credulous optimism of its own readers, had done whatever it could to make both reactions utterly predictable, since these ads—for a system that would not attain radio's state of development for some years to come—accompanied articles announcing that *Ciencia Popular* had begun building a television set in its experimental workshop. Thus *Ciencia Popular* was able to declare, probably with complete justification, that it was the first Spanish-language magazine to "conduct television experiments for the benefit of readers." It was Jorge Duclout, the magazine's editor, renowned among local amateurs, who took responsibility for building this device, which would not only receive signals but broadcast them, too.

In April 1929, *Ciencia Popular* published a photograph of Duclout's television set. It was a crude appliance fresh from a do-it-yourselfer's workbench. It had a huge wooden cabinet with a dial on top, and a tiny window 2 centimeters wide for the images produced by a perforated disk that transformed light waves into electrical pulses of variable intensity. The photoelectric tubes, familiar by now to the reading audience, had been imported for the project from the United States, and the topic of their operation inspired an analytical précis concerning the nature of the atom and the emission of electrons by such elements as lithium, sodium, rubidium, and potassium when exposed to a light source. These explanations mixed technical and scientific description into a discourse that, with all its interpolations, evasions, clarifications, predictions, and acknowledgments of limitations, was considerably more baroque than the explanations that were usually devoted to radio. It was clear from this explicative hypertrophy that a good portion of the audience had been disappointed by the outcome of the experiment, since the image quality was very blurry, the emitter and the transmitter were only 1 meter apart (so much for the wonder of *truly* long-distance communication), and the images lacked movement (a key deficit for an audience accustomed to motion pictures).

In spite of public disappointment in this device, which was not what had been dreamed of by the dozens of visitors to the special booth where it was put on display, and regardless of Duclout's primitive approach, which had more or less duplicated a radio, and despite the fundamental scientific complexity of television (radio, which was more concerned with skills and practical information, did not require anyone to understand nearly as much), the talk about television did not abate—and it was precisely here, in this discursive complexity, that the attraction to a body of technical knowledge was concentrated, since one might have a firmer hold on an icon of technical knowledge than on the aptitude for putting such knowledge to actual use. What Duclout had built was a technological icon, not a machine that could be expected to perform at the level of the thousands of radios that already existed in Buenos Aires. With radio, achieving the intended effect was a relatively simple matter, calling for nothing more than a workbench and a skilled handyman to go with it. Cinema, once it reached a mature stage of development, discouraged amateurs and turned them into audience members, not necessarily because filmmaking was technically difficult but because it was so expensive. Television was neither radio nor cinema but promised to combine them—as images viewed at a distance, as visual duplication—and make two miracles one.

At this rudimentary stage, television, like radio and cinema, rested on a new mythical effect: the technological aura, which defined the domain of modern wonders in which "wires," indispensable to the telegraph and the telephone, had disappeared[45] and broadcasting had become a mode of virtually immaterial communication. Technical explanations could do nothing to bridge the gap between the introduction of an amazing technology and the material limits it had broken through. Indeed, this was not just progress in long-distance communication but a qualitative change, which freed such communication from the material link that connected two telephonic or telegraphic devices. Obviously, television was less an actuality at this point than a potential or a hypothesis, but it had the magical aura of immateriality, even though it was heavily dependent on devices whose enormous size was out of proportion with the outcomes they produced. The spectacular potential of this innovation might be said to have been strengthened by the very absence of spectacular results.

Radio amateurs had already mastered a relatively simple set of skills, using devices whose parts could be purchased in any of several dozen shops and put together on a kitchen table. Radio, technically if not conceptually speaking, had been thoroughly incorporated into the ordinary scope of knowledge. But Duclout, king of the amateurs, busied himself with television, and his equipment was impressive even if his outcomes, which were not, created the need for articles that dealt with the attendant disappointment. Television was spoken of in the same breath as atoms, electrons, and mysterious metals like selenium—names, processes, and types of knowledge that were far removed from the ordinary practice of radio amateurs. And this, no doubt, made television much more exciting than radio, because inherent in television was the truly *fictional* element of new technologies that had acquired their technological aura as a function of keeping their distance from what could be done at home, and from the instrumentality that was more or less within everyone's reach.

The technical aura was a new phenomenon that appeared only when one zone of technology felt familiar enough for another to seem distant and unattainable. The fictional imagination, taking its place in the time lag between what had actually been incorporated into everyday life and what was still only a potential, was less interested in detailed explanations of processes than in the story of what those processes would make possible once they had been completely mastered. Television reintroduced a gap, a distance to be covered, between technology and its users: despite the most optimistic forecasts, television was still *not quite here*; thus, to the extent that it ran ahead of the expectations of those who had already adopted radio as an everyday cultural reality, television carried the promise of a future. By showing that there were still unimaginable feats up ahead, television appropriated the wondrous nature of radio's origins. And even though the introduction of radio was still relatively recent history, it seemed, not without reason, to have been superseded by the strides radio had made since then, and by how rapidly it had spread to every level of society.

With television, at any rate, radio amateurs were forced once again to confront the problematic fact of information and know-how that surpassed their knowledge and technical skills. Duclout's outsize television set was a disappointment to many people precisely because they had expected more, had expected everything: long-distance cinema, with sound and movement and sharp images and defined contours and luminosity, all on a large screen. That had been the promise in those days, and Duclout's device shrank that promise down to the dimensions of an unfinished rough draft. But the future was hiding in this disappointment, because the domain of what had not yet become possible contained a privileged space for technological fantasy, a place that magnetized desire and provoked daydreams, if not understanding. Over the course of a few short years, the pioneers of radio had caused the number of radio sets to multiply, but television showed them the limits of their knowledge and skills. For precisely this reason, however, television was a present impossibility that waited in the possible future. No one knew then that it would take almost a quarter of a century for that future to arrive.

The continuum running from the phonograph through radio and cinema to television had started from what actually existed, and it was drawn toward what did not yet exist as a real possibility within the technological framework of the era. Both poles of the continuum were necessary: what did not yet exist opened up the dimension of technological wonder (of the technological aura), and everyday practice rested on what was already possible. Within a very short period, radiomania had synthesized what was yet to be with what was already possible, offering proof that technological desires and fantasies could produce real effects on ordinary life. The time it took for television to come into being did not contradict this experience but did delay its realization, although the experience was nourished by a confidence that had not yet been shaken. This blend of future possibilities with present attainments gave rise to a dimension in which technological reasoning reinforced imaginative expectations. And, as we know, there was nothing more powerful than the synthesis of current limitations with wonders yet to come, in which the fulfillment of the technical imagination already seemed to be a fact of everyday experience, an experience, it is worth emphasizing, that was inclusive from a social and democratic standpoint because of its relative independence from institutionalized learning. The autodidacts and the natural mechanics were treading familiar ground.

CHAPTER SIX

Doctors, Clairvoyants, and Quacks

> These hallucinations, which are sometimes compelling, lead in so many cases to criminality that Dr. Krafft-Ebing has divided crimes into acts of melancholia having one of three causes: feelings of sadness, periods of anxiety, and hallucinatory phenomena. The latter, according to Wilhelm Griesinger, can result from deep mental or physical exhaustion.
>
> —Roberto Arlt

This chapter's epigraph was inscribed in the space of a fantastic psychology, both old and new, which had been stitched together from scraps of the Spiritism, hypnotism, and clairvoyance that accompanied, like a ghost, the popular interest in technology, scientific curiosities, and fabulous discoveries. Old obsessions of traditional culture—from tales of "apparitions" to *curanderismo*, or folk healing, in the countryside—met a new urban discourse in the form of theosophy, parapsychology, and popularized psychological and even psychoanalytical ideas.[1] A new mythology was making selective use of a pseudoscientific lexicon to translate such persistent obsessions as communication with the hereafter, life after death, the fountain of youth, the desire for miracle cures, and thought transference. These topics had also been favorites of the popular imagination before the twentieth century, but now, combining in highly unexpected ways with the novelties that were being introduced as scientific advances, they shaped a mythological domain that persists in our modern-day press. In the 1920s, with the consolidation of a mass-circulation press, such topics found an area of public space that was larger than that of tales and oral tradition.

There was no reason not to believe that everything was possible. The speed with which some technological changes had come to belong to everyday life lent support to the notion of miracles while inspiring well-known metaphors—the miracle of electricity, the miracle of radio, the miracle of cinema. These and other modern wonders comprised a cultural landscape where every promise could be fulfilled. Radio and wireless telegraphy had offered spectacular proof that material obstacles could be overcome, and that an epoch had begun in which perception would no longer be ruled by the corporeal limitations of the senses or the physical limitations of their extensions (spectacles, magnifying glasses, telescopes, and the like). And now that sounds and images were being broadcast by way of invisible, disembodied channels, it was possible to construct a whole system of correspondences regarding other types of long-distance transmission and reception. Moreover, if X rays could penetrate the body in a way that had never before been possible without a scalpel, and if the inside of the body could be photographed and its material envelope rendered invisible, then science was staging "miracles" that authorized a belief in miracles of other kinds.

And yet such changes had come about in much less time than it takes to live a human life. People who were adults in 1920 could remember a childhood without electric lights, telephones, radios, movies, or sophisticated medical devices. These developments pointed to a great technological and scientific turning point, which had taken place in their own lifetimes but quickly became ordinary (the adaptation to technological advances in general living conditions occurs with unparalleled speed); at the same time, however, as far as do-it-yourself technology was concerned, there was no diminishment in the sense of wonder. The technological advances were on the scene, and even though thousands of amateurs in the areas of mechanics, radio, and cinema were using them in a way that was more or less informed, others had no interest at all in understanding the scientific principles that had made these advances possible. Whether or not one knew *why* things worked the way they did, the fact was that anyone could *make* them work. Knowledge that had once been indispensable, if one wanted to enter the world of long-distance communication, cinema, or automobiles, had dwindled down to a bare minimum. For many people, the sense of technological wonder had quickened the desire to learn the new laws of this progress that prevailed by defying common sense and validating old myths, such as those involving the ability to talk with a person who was not physically there or hear the voice of someone who had died or see a picture move, not to mention the body itself giving up its insides without losing so much as a single drop of blood. Others simply surrendered to the sense of wonder, without taking on the challenge of acquiring mastery or information. But they were all pulled onward by the frenzy of the new, considered synonymous with indefinite, unlimited innovation. Radio and cinema had come along, and so could any other type of long-distance communication. If flight was possible, then trips to the moon were right around the corner. If the body could be opened up by invisible light rays, then it would also yield to other, equally surprising technologies. The more spectacular the technological application, the stronger the belief it engendered in its infinite developmental potential. But these novel forms of newness were also evoking other, far older ideas—clairvoyance, the profound gaze of the fortune-teller or *curandero*, miracles—and if real miracles had already occurred, then why couldn't there be more, no matter how implausible they seemed?

In a way that only appeared contradictory, technological novelty failed to produce skepticism about wonders, miracles, and superstitions; instead, it created openings to these earlier forms of belief and practice. If the evolution of technological progress knew no bounds, then an expanded system of corresponding possibilities could also be imagined, through a form of reasoning that operated by analogy, as much of science fiction did:

> We have now become completely familiar with radio broadcasts of music and speech, and we are confident that before long we will have television to enjoy as well. Further advances in radio-electronic science may help to strengthen the bonds that unite all men, in every country. There are some who promise that the final, most extraordinary invention will be direct thought transmission from a master brain to the multitude, by means of an exceptionally powerful electromagnetic wave broadcasting ideas at lightning speed.[2]

This was precisely the fictional premise envisioned by Aldous Huxley, George Orwell, and Yevgeny Zamiatin; and the author of the article just cited was afraid, as they also were, of this premise coming to pass in a reality that was still in the future, but not by much. And yet what also leaps out here is the specific potential presented by the immaterial nature of a radio broadcast. In retrospect, it is clear that radio created conditions in which fantasies of clairvoyance and disembodied communication, including messages from the hereafter, were completely plausible. Thus, for millions of readers and listeners, the leading edge of the modern served not only to reinforce the appeal of technology and technological information but also to strengthen beliefs in phenomena that themselves depended on the capacity for an infinitely open, ever-expanding astonishment. And herein lay an ambivalence, which was obvious in how newspapers and magazines treated this kind of news, and just as obvious in the blend of fascination and denunciation that ran through Roberto Arlt's text on the "occult sciences" in the city of Buenos Aires.

Technological progress fed new fantasies, which coexisted with the concrete skills needed to operate machinery. Wonders had not ceased; they were reintroduced instead, in the guise of new materials and new experiences. On the one hand, technological wonders and modern miracles endowed technology with the qualities of fiction; on the other, they enabled reasoning by analogy where the wonders of earlier times were concerned, repositioning them and sometimes giving them new respectability. And all this ancient wisdom was being recycled in a democratic way—if there was as much hypnotic power being exercised in psychiatrists' offices as there was in theaters,[3] the hypnotic techniques disseminated by newspapers and magazines acknowledged no limitations that could not be transcended by an amateur who could summon the necessary patience. Correspondence courses were marketed over the radio, newspapers like *Crítica* featured methodical lessons, and there were technical dictionaries in a number of other publications, but courses in self-improvement and self-hypnotism were being offered, too, and *Ciencia Popular* published a complete do-it-yourself hypnotism manual in five installments.[4] The first lesson focused on general principles, emphasizing precautions, recommendations to follow the steps in their prescribed order and avoid setting goals too difficult for a beginner, and other commonsense advice. What was most significant, however, was the firmness of the manual's assurances that hypnotism was a technique (the manual defined it as, strictly speaking, a "branch of science"); and, that being the case, not only could anyone become a skilled practitioner, "higher learning" was not even required in order for one to "accomplish surprising things" with hypnotism. This is how it was possible to democratize a power that evoked, at one extreme, the immateriality of magic and, at the other, a turn-of-the-century psychiatrist's office (equally inaccessible to the readers of *Ciencia Popular*), and even though no one actually became a hypnotist by studying this manual, its publication took a phenomenon that might have been considered a product of special powers and gave it all the transparency of something that could be taught. At the same time, because this manual appeared in a magazine that was devoted almost exclusively to new technologies, old-school hypnotism found itself recycled within a framework of modernity. Thus boundaries were blurred on two fronts—technological miracles were "modern wonders" conferring plausibility on impossible feats that were now assumed to be possible, or on a distant future that appeared to be an almost direct continuation of the present. Moreover, the notion of hypnotic technique as teachable know-how (witness the pedagogical efforts expended by newspapers and magazines, correspondence courses, and plebeian training institutes) conferred the same teachability on more mysterious bodies of knowledge that were anchored in personal powers, and these too were now presented, at one blow, as skills that could be acquired with study and practice. Anybody could perform hypnosis, and so the point was for practicing hypnotists not to make a secret of their perfectly teachable body of knowledge, thereby cheating those who had not yet discovered that everyone possessed the power of suggestion. There were forces both invisible and real, and there were more of them than people knew. This lesson from science had a mythic (and poetic) power that called everything into question for almost everyone, and the various discourses that came together under its banner had quite disparate relationships to experimentation and knowledge.

In 1922, *Crítica* published a translation of an article about the "invisible world" by the French astronomer and best-selling author Camille Flammarion, a new convert to Spiritism, like another great thinker, Arthur Conan Doyle.[5] The text expounded a conviction, as trite as it was undebatable, that lent respectability to every sort of parapsychological or Spiritist fantasy:

> There are more unknown forces than known ones, the sciences are at the dawn of their development, and what we know represents the tiniest of islands in the middle of an ocean of unknowns and superficial appearances. For the past twenty-five years, unexpected discoveries in physics and occultism have been causing us to divine the existence of unobserved panoramas that nowadays are accessible to our spirits, which for so many centuries were sound asleep on the pillow of indifference.[6]

Who was prepared to take issue with this generalization about a universe whose extent, properties, and laws were still not completely known? From this obvious statement, Flammarion moved on to defend the unknown dimension, speaking of it in the tones of scientific positivism. Thus he established a basic system of correspondences: physical phenomena were studied, and so psychic phenomena had to be studied, too, and they would certainly be found to include instances of thought transmission that circumvented material ties, just as wireless telegraphy had already proved that long-distance disembodied communication was possible. This analogy allowed Flammarion to elaborate on the possibility that communication could be established between the living and the dead, since the living unquestionably communicated among themselves. And, Flammarion continued, events that the scientific positivism of fifty years ago had judged to be illusions were now being studied and analyzed by the "science of today" and would be fully explained by the "science of the future." Most of Flammarion's "scientific" explanations of unusual phenomena were based on this analogical epistemology. Thus a road that had started from an expanded system of comparisons led to a relatively simple, understandable mechanics, which took radio as its own point of departure in asserting that messages were pouring in from other dimensions of time and space.

Reasoning along with Flammarion, one could devise analogous explanations for most of the phenomena discussed at Spiritist meetings and in Spiritist publications. For mass-circulation newspapers like *Crítica* and *El Mundo*, the Spiritists themselves fell neatly into one of two camps—liars pretending to be mediums, or true theoreticians and practitioners of disembodied communication.[7] Spiritism had been practiced in Buenos Aires at least since 1877, when Angel Scarnicchia founded the Sociedad Constancia, which soon launched a magazine, *Constancia*, edited by Cosme Mariño, the most eminent of the Argentine Spiritists, who was always treated respectfully by the Buenos Aires papers, where he was described as the Río de la Plata region's answer to Allan Kardec.[8] This was fairly predictable, given the context of a press on the lookout for striking news. More surprising is the fact that, in 1926, the department of philosophy and letters at the Universidad de Buenos Aires offered a platform to a lecturer who refuted local and international Spiritism's most extreme ideas and denounced the deceptions of mediums but still left room for the hope that science might one day manage to "shed light on shadows full of mystery." In fact, it was those shadows that formed the basis of the fantastic psychology and parapsychology that were the subject of so much reasoning by analogy in Spiritist pamphlets, but not in them alone. News from abroad, which reverberated in Buenos Aires, was reporting scientifically controlled experiments in parapsychology, although these usually ended in scandal when the medium was shown to be a fraud or, at best, a hysteric. But this kind of news did spur public discussion among scientists, especially when the articles were about conversions as sudden and unexpected as those of celebrated figures like Conan Doyle, or when an article dropped names, such as those of the Institut Métaphysique de Paris or the Congrès Internationale de Sciences Psychiques (held at the Sorbonne in 1927), that evoked prestige and an aura of reliability.[9]

At any rate, in the same context where one read about instances of technological progress that seemed literally impossible, this news was also being published, sometimes with ambivalence and sometimes with the aim of investigating the true nature of hitherto unknown phenomena that were as mysterious as magnetism or electricity had been in the eighteenth century. Within the scope of this discourse, a feast of sensational novelties was laid out before the public, combining news about disembodied phenomena with dissemination of the Spiritist principles that were being promoted by the local sects.[10]

The big question posed by such articles had to do with the nature of the elements in play. Were they indeed natural but invisible forces, previously unknown but belonging to the order of mineral magnetism or animal electricity?[11] Or were they supernatural, caused by "spiritual entities"? The issue, and the true Achilles' heel for a science of the invisible (if such a thing could ever be possible), was control over the experiments that were conducted with mediums. A tone of skepticism was blended with a tendency to swallow what were actually dubious proofs, and for every mediumistic experiment that could be shown to have been scientifically controlled, another was reported to be a fraud. This mass of published articles was as heterogeneous as its sources in the news; overall, however, the prevailing impression was of something going on in that little-known dimension of spiritual, animistic, or astral relationships, and if there was fraud and self-deception to spare, some experiments also reported events that still could not really be explained. These phenomena needed a science that could study them, both so they could be explained (as it had finally been possible to do with magnetism) and so objective experimentation could be separated from religious belief, on the one hand, and charlatanism, on the other.

As a parallel activity, the newspapers took it upon themselves to denounce the frauds promoted by false clairvoyants and bogus fortune-tellers, drawing a line of differentiation that separated parapsychology and "respectable" Spiritism from the impostors who faked spirit manifestations and promised communication with the dead (prepaid, of course). The exposé at *Crítica* coincided with a crusade being waged by the local authorities, who were going out to the neighborhoods throughout 1926 and 1927 and raiding the homes of those suspected of being petty swindlers.[12] But the same papers, with no excess of irony, also reported on drop-in visits from clairvoyants and mediums. Sometimes the reporter writing such an article would mount a ferocious verbal attack on the visitor, but the visitors kept on coming, and the articles kept appearing in a steady but irregular stream, often showing extreme credulity in the face of a curious amalgam of scientific experimentation and theosophical rambling that moved back and forth between addressing a people's tribunal (the audience) and one made up of specialists:

> "My experiments are surprising," Professor Bernardo explains, "because they reach people, without reliance on tricks of any kind. These are the same experiments conducted in hospital lecture halls or medical clinics by scholars like Charles Richet, and by students of the unknown, like the recently deceased writer Arthur Conan Doyle, and, in the past, by the eminent Doctor Jean-Martin Charcot. Like my Indian spiritual guides, I conduct these experiments before the public, explaining the scientific principles involved. My experiments are an exhibition of those primordial things that form the basis of the religions—the things that form the basis of all of modern Esoterica. Skeptics are always the first to become my followers. As the Buenos Aires public comes to know about these scientific tests, controversy will arise, and this interest is certain to have wide repercussions in scientific circles."[13]

Like a stage hypnotist, Professor Bernardo deployed his science in a theater, where the body of his medium was electrified so that mere contact with it was enough to light dozens of lamps or scorch a human being. There were also the usual demonstrations of hypnotism and thought transference, including the well-known gambit in which a member of the audience gives the medium a silent command to perform a tango or a classical piece of music. But while this hypnotic circus was exercising its powers of fascination, it was also offering guarantees of its authenticity to spectators who were probably in a position to control every aspect of the experiment. Seamlessly, and within a single paragraph, the reader is taken from Charcot to what was practically a carnival act. This kind of contiguity was characteristic of the journalistic tone (and of the popular tone it probably inspired) that was used in connection with clairvoyance, parapsychological forces, and hypnotism, each of which actually operated by its own very distinct rules. Nevertheless, under the rubric of wonders that had once been mysteries to science but were now beginning to find objective clarification, there was a little of everything, with no need on the part of journalists (or readers, perhaps) to draw any lines of demarcation other than the thin one that already separated mere trickery from outright fraud. The obscure but promising zone of psychological wonders had room for a fantastic psychology, which still awaited rational explanation.

Weird Science

In Leningrad, a team of scientists decapitated a dog and, keeping the dog's blood circulating outside its body, managed to keep its head alive, as in a science fiction story:

> At one point the mouth attempted to open, as if, it was noted, it wanted to bite, the physical reactions of an agitated dog being observed whenever the dog was greatly provoked. The dog—or, better, its detached head—responded with almost all the normal reflexes. . . . Several drops of water were dropped onto the dog's tongue. The head immediately displayed agitation, and the tongue made movements to expel saliva. . . . Sweets and bits of cheese were given to the animal, emerging in pieces from the esophagus as soon as they had been ingested.[14]

At issue here was not just whether it was possible to implement the technique of extracorporeal blood circulation, which would actually come into use later on, in operations performed during the second half of the twentieth century. There was also the question of the line between life and death, and of the possibilities for resurrection outside a religious context. If it was possible to look at a photograph of one of these scientists staring into the dog's eyes (eyes that were, the caption promised, just as alive as the scientist's), then nothing could stand in the way of the fantasy that life continued beyond apparent death. Questions that were perfectly scientific in one domain were mythic in another, and contamination between the two was unavoidable in the communication that newspapers and magazines established with astounded readers who were assured, in short, that improvements in equipment and procedures would eliminate the technical problems that had plagued this first experiment.

If the only limit on scientific progress was time, and if technology alone was enough to impede or liberate knowledge, then a public convinced that progress was both imminent and inevitable was prepared to accept all the "miracles" that an evolving science could go on working. It was not unusual to read news about experiments that were shocking in their breach of moral and technological boundaries, and that evoked science fictional fantasies resembling the experiments depicted by H. G. Wells at the end of the nineteenth century in *The Island of Dr. Moreau.*[15] *Crítica*, for example, reported in October 1927 that scientists in Soviet Turkestan had attempted to cross monkeys with humans in order to test a Darwin-inspired system of selection, for which natural selection would be speeded up in the laboratory through the manufacture of two biological lines, one of human-monkeys and the other of monkey-humans, both conceived by artificially inseminated female monkeys.[16] Nine months earlier, readers of the same newspaper had confronted the unbelievable and unnerving news that a chemistry lab in Berkeley, California, had succeeded in creating life. This article was somewhat more respectable from a medical standpoint, as much because of the way it described the experiment as because of the fact that the aim in this instance of genetic engineering had not been to create semihuman beings; its fictional effect was just as striking, however.[17] Everything seemed possible in a science open to the Faustian dimension that was challenging the fictional and philosophical imagination. For the first time, a mass audience was seeing, right before its eyes, miracles that had not come from the world of myth and religion but instead from what was presented as the province of science, where disturbing investigations were being carried out and crossing a line that only magic and fantastic fiction had ever come near. But since the results of these experiments could not be verified, and since the procedures for verification were different for different experiments, the fact that scientific novelties like these had actually been shown to be possible was not what mattered. What mattered instead was that these novelties, which did nothing more than create an illusion of modern miracles, made science into something that was "interesting" to a public wider than the educated segment of the audience. Science was now linked with questions and desires that involved the origins of life and the power to create—what had once been the property of the gods was under new management.

And the miracle cures of folk healing, along with religion and magic, had come just as far as science. The Soviet scientists' golem, a terrifying cross between a monkey and a human being, had greatly pushed the limits of what was possible, strengthening rather than weakening belief in a science that was presented as making the failed dreams of wise men and magicians come true. Science promised not only spectacular progress in ordinary life but also the literal opposite, a future marked by events that were all but supernatural. Regarding the latter, it was difficult to tell sensationalism from objectivity, but that was certainly of no importance when it came to science casting its spell over the collective imagination. Scientific fantasy—which journalism disseminated as one element in a mixture of news that was likely, or possible, or actual, as the case might be—found a place for itself where it would be welcome as long as other, earlier beliefs were not driven out. All these discourses intersected, seeming to reinforce one another in speaking of the same things.

Miracles and Frauds

> One among many hundreds of foreigners walked into our editorial office and introduced himself as a "doctor of psychology" specializing in "sun treatments."
>
> "I'm German," he said. "I've spent five years in Argentina, and have just arrived from the province of Entre Ríos. I studied medicine at several schools in Germany but didn't graduate, and am not a medical doctor. . . . I've discovered that I have unusual natural gifts. My spirit is in intimate contact with the spirit of the sun, and all I have to do is wish for the sun to use me as its channel and direct its powerful rays to any ailing part of a patient's body. I give off white, gold, and violet rays and can destroy any disorder of the human body. If I focus on a tubercular lung, the organ is transformed."[18]

The German cited here used a blend of transcendentalist and "scientific" lexica to describe his powers, seizing upon the medical-sounding word "rays," and his knowledge appears to have belonged to a modernized variety of folk healing. But there were still *curanderos* of the old kind, and they were the targets of hard-hitting journalistic exposés. In 1923, *Crítica* dealt with this topic almost daily, publishing names and addresses as well as facts, and directly questioning the police, accusing them of a suspicious lack of action regarding an area of criminal behavior that was a hotbed of Spiritists, clairvoyants, and folk healers known as *manosantos* and *manosantas* along with prostitutes and pimps. The bookshop and novelty store at 2532 Calle Oro offers a succinct example of this multiplicity of trades. It was a "den of deceptions" that encompassed D'Adda ointment, for "breast development"; Apletonia, a remedy for baldness; relics and sacred images; and wild plants whose descriptions were provided in a catalog that the two women who owned the store offered to mail to whoever requested it. Another of the building's tenants, in league with these two businesswomen, was running an operation that employed "several women of ill repute."[19] The cost of consulting a healer or a clairvoyant was generally 1 or 2 pesos, and their advertising targeted a clientele primarily made up of women so devoted to these unconventional procedures that, six years after the exposés, newspapers in Buenos Aires were in agreement that the city was being "undermined by *curanderas*, *curanderos*, and fortune-tellers" who were "taking advantage of the public's gullibility." On the basis of their skills and techniques, these practitioners were divided into various categories. There were *manosantas*; evangelists for "water cures"; naturalists; and proponents of hot baths, cold baths, and steam. There were quack physicians and dentists who set up shop in offices where they charged less than licensed professionals, and there were masseuses specializing in gout, sciatica, stomach and kidney ailments, and nervous disorders. There were also Spiritists and "somnambulators" who dealt with psychological matters, specializing in love troubles.[20]

The *curanderos* were associated with a paramedical art linked to a world of tradition that had its own forms of illusion, belief, efficacy, and fraud, but the truly electrifying miracles were being introduced from the scientific quarter or, better, from the quarter of whatever was able to cloak itself in scientific discourse, or resort to the lexicon of science, or be expounded from scientific lecterns, or use the guidelines of medical practice as an organizing principle. And in the second half of the 1920s, two such techniques were galvanizing the public imagination and exciting the hopes of thousands of people: the grafts of Dr. Serge Voronoff and the treatments of Dr. Fernando Asuero.

The Voronoff method was first used in Argentina at the Hospital Piñero in April 1926, when Dr. Richard Spurr operated on a young man of seventeen who had "rudimentary mental faculties" and suffered from complete atrophy of the gonads.[21] This was a procedure for grafting chimpanzee gonads into those of a human host. The chimp, placed under a bell jar into which openings had been made to accommodate the contents of eight ampoules of chloroform, was then transferred to an operating theater "packed with physicians and practitioners" who witnessed the intraglandular graft that Spurr, assisted by two other surgeons, carried out on his patient. Within a few days, the boy, who before the operation "had not been able to say just what he had been doing only hours earlier," was able to recall "even the most minor details about himself" and, what was "most remarkable," could feel "sensations and phenomena that he had never before experienced." Three months later, this lucky patient was working in the hospital's gardens.

Spurr was in the headlines for weeks, not only because he had allegedly achieved a triumph of medicine but also because, in order to do so, he had taken the liberty of altering Voronoff's method. Instead of stitching "a portion of chimpanzee gonad, cut into very thin slices, into the gonad of the surgical patient," he had implanted the entire gland, using careful incisions and sutures to avoid the postoperative pain of the inventor's patients, generally sexagenarians who, in search of a rejuvenation therapy, had turned themselves over to Voronoff or his followers in European clinics. Months and years later, news about catastrophic results—or, in most cases, no results at all—began to leak out of these clinics, including one in Vienna. But other information, which gave assurances that these operations had been completely successful, was published right alongside this news, and the Vienna board of physicians heard testimony from the method's first two beneficiaries.

In spite of the debate, Spurr was still grafting gonads two years after his first operation, even though by now it had become necessary to tack 500 pesos onto the standard cost of the procedure so as to cover the expense of procuring a suitable chimp. Spurr expressed his hope that a chimpanzee-breeding industry would get under way in Argentina and make it possible to treat not just people who were sick but old men as well, to whom this type of graft would restore lost vitality, especially sexual vitality. Up to that point, the middle of 1928, only the Buenos Aires zoo had been providing the indispensable chimpanzees, and the zoo's supply was not unlimited.

Enrique Dickmann, among other eminent physicians, sharply denounced as frauds all those who were performing Voronoff grafts in Argentina, but the daily papers preferred to continue down the path that the wire services had cleared with their cables celebrating Voronoff's appearances at this or that spot around the globe. "Some aspects of modern medicine," according to Dickmann, seemed "more like witchcraft than science," but they were also much more challenging to the imagination than was the cautious skepticism of this critic of Voronoff and Spurr. Dickmann, like a good socialist, proposed to combat the irreversible passing of time with a regimen based on teetotalism and physical exercise. This "American therapy," as Dickmann himself called it, lacked the instantaneous magical and mythic power not only of Dr. Spurr's treatment[22] but also of the rejuvenating serum produced by Doctor Josef Franz Kapp from a precipitate of pigskin that was injected into rabbits whose tongues, after two weeks, were extracted, placed in a container, and separated into a clearer fluid and a darker one. The clear substance was the base of an elixir that, in this case, would mitigate the effects of old age in women, whose anatomy made them ineligible for the Voronoff treatment.[23]

In May 1930, who should arrive in Buenos Aires but the other hero of this Faustian mythology that had conquered the fringes of science: Dr. Fernando Asuero, a Basque physician whose method was based on cauterization of the trigeminal nerve, a simple operation with which Asuero claimed to have cured hundreds of different ailments. His arrival provoked a philosophical disquisition from *El Mundo* that was fiercely critical of scientific positivism:

> Nothing about the twentieth century is positive. Everything has been reduced to the unending investigation and acceptance of facts, whatever their sources may be. The British have set a meaningful example in embarking on what they claim will be a serious scientific study of spiritual and after-death phenomena—a claim that certainly must have scandalized any remaining followers of Auguste Comte, who do not believe in any possibilities for science other than those offered by the academy. In this sense, Asuero is a man of our time. Welcoming the academic onslaught with an ironic smile, he continues to employ the method that is indisputably his own, regardless of any precedents that others may wish to find in it. His tough Basque spirit has little interest in what official science has to say about his system, as long as patients, even a few, are finding relief in trigeminal cauterization.[24]

The first opposition to be noted here is the one between positivist science and a new science sensitive to phenomena that positivism had barred from its field of study. There was more to the physical world than what our senses grasped; mental phenomena, life after death, and resurrection could be objects of study as legitimate as any others. A second opposition pits a living, open-minded science against the academic science that was the refuge of nineteenth-century positivism. In science, to be modern meant to accept the sometimes intangible and strange nature of phenomena that actually did exist, even though their sources could not be explained in traditional cause-and-effect terms. The science that was evolving outside the academy was experimenting open-mindedly and dynamically with the effects of its procedures—it produced cures, even though it could not completely explain them.

El Mundo's self-styled "epistemological" introduction to Asuero's visit to Buenos Aires was followed two months later by another intercession on behalf of Asuero, this one even more valuable as a gauge of the general atmosphere because it came from someone who was known for his corrosive skepticism and his denunciations of the local *curanderismo* in its usual forms. Roberto Arlt, in one of his essays on life in Buenos Aires, defended Asuero's method because of his own opposition to the remote, skeptical medicine practiced by the doctors who were sacred cows on the local scene, the "nouveaux riches of science" who all had "the same toupee, the same hunger for recognition, and the same air of infallibility."[25] Arlt contrasted the business of official medicine with a method that aroused feelings of rivalry in the established practitioners precisely because it was inexpensive and had immediately verifiable results, thus posing an obstacle to the prolonged treatments by which doctors kept their patients bound to them until the patients died. Here again, Arlt was pointing to the issue of control over a body of academic knowledge on the part of people who had been left out, and who now made up a legion of the defrauded. With Asuero's method, they regained control over their treatment, obstructing the "traffic in sickness" and cutting off a source of fees—and fortunes.

El Mundo had welcomed Asuero in April 1930, a month before he began his treatments in Buenos Aires, and these came as a surprise to the press and the medical community (or at least that portion of it surrounding him), not just because of his experiments in hospitals but also because of the lavish attention showered on

Asuero at innumerable crowded lectures, on excursions by yacht down the Paraná Delta, and during strolls to the racetrack, where the doctor was greeted in the paddock with polite applause.[26] As Arlt's essay had predicted, there were dissident voices or shades of distrust from doctors demanding that Asuero present his successful cases before their respective specialty boards, since some doctors in Buenos Aires, even before Asuero arrived, had also been offering trigeminal-nerve treatments.[27]

A simple, practical demonstration carried out at the Hotel España before several Argentine doctors, and performed on a friend rather than a patient, brought Asuero front-page coverage in Buenos Aires:

> The Basque physician placed three probes into the subject's nostril, inserting the first up to the pharynx, the second up to the middle concha, and the third up to the inferior concha. The effects of this procedure were just as Dr. Asuero had described them beforehand, and the demonstration came to an end.[28]

The mundane nature of the setting, the freedom from tension (because the method had not been used on a patient), and the schedule of activities and invitations that preceded and followed every one of Asuero's simple, clear experiments created an atmosphere that, although it was public, was nevertheless separate from the drab environment of the hospital, the narrow specializations of board-certified physicians, and the pain or anxiety of people hoping for a cure. Asuero offered amazing but "neat" treatments whose quickness guaranteed their efficacy, or at least ensured that his contact with the trigeminal nerve would not be followed by days of a painful convalescence with uncertain results and consequences. Dr. Asuero—a bon vivant, as the news reports described him—had found a therapy that, with its speed, and given the circumstances in which its steps and effects were demonstrated, was almost theatrical. The press made his practices accessible to the masses (as audiovisual media would be doing in our own day), and he dispensed with bloody operating rooms and the long-drawn-out periods needed for conventional treatments. At the same time, however, his practices were physical; he touched the body, thus disarming the skepticism of those who, like Arlt, denounced long-distance treatments and parapsychological interventions. Asuero's method, although alien to the "followers of Auguste Comte" (as *El Mundo* described the scientific positivists), adopted the conventions of the most commonplace medical treatments. It was quick, bloodless surgery that forged a solid pact between the desire for good health and the distrust of official medicine. The memory of Asuero persisted until the 1950s in the collective imagination, which was still, twenty years after his visit to Buenos Aires, attributing impossible cures to a simple nudge of the trigeminal nerve.

L'affaire Asuero ended predictably, and very quickly. There were denunciations coming from Spain, and local disrepute ensued when patients who had been miraculously cured began to deteriorate. One of his followers suffered the misfortune of having a patient die, and Asuero, now enthusiastically criticized by the same papers that had been lionizing him a few weeks before, made what was practically a getaway to Chapadmalal.[29] Not long afterward, the health department pressed charges, and in June a judge served Asuero with a writ of subpoena. And after these final news items, the trail goes dead.[30]

The parable of Asuero is quite typical of how the mass-circulation papers processed the raw material of the news. A variety of up-to-date scientific information, including *Crítica*'s 1926 survey of medical science in Argentina,[31] alternated with the pursuit of medical oddities and, on a parallel track, miracle cures. At any rate, both Voronoff and Asuero, who had been wrapped up in the hopes of the public in the most curious ways (Orfilia Rico, a theatrical luminary of those years, confided that she had been treated with trigeminal therapy), also appear to have been harassed by those who denounced them as impostors.[32]

What to make of all this? Pieces of confirmed, likely, unusual, and implausible information, arrayed against the backdrop of a tendency to believe that anything was possible, were reinforcing one another instead of canceling one another out. If ordinary people continued to rely on *curanderos* and clairvoyants, they also transferred their faith to the miracle cures of Voronoff and Asuero, and denials or denunciations carried little weight next to the wish to believe that a universal cure-all or Faustian secret was only a step away. But the credulity of the journalists and readers who celebrated these medical miracles was fed by more than myth, and more than the past. There were modern miracles of technology and science, too, and imagination blended the past with the future, admitting no sharp separations but rather a continuum on which the power of the new, which had already been confirmed, sustained the belief that impossible things were indeed possible. People in search of youth, health, and eternal life found strange encouragement for their desires in the realities of scientific progress, but the irony was lost on them.

Notes

Notes

Introduction

1. Beatriz Sarlo, *Una modernidad periférica: Buenos Aires 1920 y 1930* (Buenos Aires: Nueva Visión, 1988).

2. [Leopoldo Lugones, *Las fuerzas extrañas*, ed. Pedro Luis Barcia (Buenos Aires: Ediciones del 80, 1987, orig. 1906).—Trans.]

3. Roberto Tálice, *Cien mil ejemplares por hora: Memorias de un redactor de Crítica, el diario de Botana* (Buenos Aires: Corregidor, 1977), 280–81. [Natalio Félix Botana, born in Uruguay in 1888, was the publisher of *Crítica*.—Trans.]

4. One aspect of the theoretical issues surrounding the social imagination or imaginary, including a discussion of recent literature, can be found in Carlos Altamirano, "Lo imaginario como campo del análisis histórica y social," *Punto de Vista* 38 (Oct. 1990). ["The social imagination or imaginary" ("imaginación o imaginario social"): an apparent allusion to Jacques Lacan, who posited the imaginary as one of three orders making up subjectivity, the others being the symbolic and the real.—Trans.]

5. There was nothing unusual about this formative capacity of scientific or technical disciplines incorporated into everyday life. See, for example, Lynn Merrill, *The Romance of Victorian Natural History* (New York: Oxford University Press, 1989), on the reading of natural history during the Victorian era as a way of instilling good habits in the lower orders, and on its power to generate motifs and images of collective identification. For other nonscientific functions of scientific knowledge, and for technical ideologies, see Thomas Hughes, *American Genesis: A Century of Invention and Technological Enthusiasm, 1870–1970* (New York: Viking, 1989); Stephen Kern, *The Culture of Time and Space, 1880–1918* (Cambridge, Mass: Harvard University Press, 1983); and Cecilia Tichi, *Shifting Gears: Technics, Literature, Culture in Modernist America* (Chapel Hill: University of North Carolina Press, 1987). On intellectuals and technological progress, see Leo Marx, *The*

Pilot and the Passenger: Essays on Literature, Technics, and Culture in the United States (New York: Oxford University Press, 1988); Leo Marx, *The Machine in the Garden: Technics and the Pastoral Ideal in America* (New York: Oxford University Press, 1964); and Flora Süssekind, *Cinematograph of Words: Literature, Technique, and Modernization in Brazil,* trans. Paulo Henriques Britto (Stanford, Calif.: Stanford University Press, 1997). On the ideologies of progress and their relation to history, see Peter Bowler, *The Invention of Progress: The Victorians and the Past* (Oxford: Blackwell, 1989).

6. The mythic impact of the new technologies was cogently demonstrated by Jorge F. Liernur and Graciela Silvestri on page 4 of a manuscript that originally had the working title "Para un estudio de las relaciones entre tecnología y cultura, Buenos Aires, 1880–1930: El torbellino de la electrificación," eventually published as *El umbral de la metrópolis: Transformaciones técnicas y cultura en la modernización de Buenos Aires, 1870–1930* (Buenos Aires: Editorial Sudamericana, 1993). Where the introduction of technology is concerned, their work deals with a history that follows paths very similar to those taken by the present volume in its theoretical reconstruction of the popular imagination's technical dimension. In their original manuscript, Liernur and Silvestri also note the contrast, in the first decade of the twentieth century, between advances in technical knowledge and the persistence of outdated scientific notions, even in manuals published by academic authorities.

7. The work of several writers offers support for theories about the density of pop-culture networks; see, for example, Luis Alberto Romero and Leandro Gutiérrez, "Sociedades barriales, bibliotecas populares y cultura de los sectores populares: Buenos Aires, 1920–1935," *Desarrollo Económico* 109 (Dec. 1989).

8. On cures brought about through hypnosis, magnetism, and parapsychology, see Robert Darnton, *Mesmerism and the End of Enlightenment in France* (Cambridge, Mass: Harvard University Press, 1968).

9. See Adolfo Prieto, *El criollismo en la formación de la Argentina moderna* (Buenos Aires: Sudamericana, 1989).

10. On the importance of strategies for assimilation through institutional and noninstitutional means, and their resulting variations, see Raymond Williams, "Base and Superstructure in Marxist Cultural Theory," in *Problems in Materialism and Culture* (London: Verso, 1980), 39, 40: "We can only understand an effective dominant culture if we understand the real social process on which it depends: I mean the process of incorporation. The modes of incorporation are of great social significance. . . . The educational institutions are usually the main agencies of the transmission of a . . . culture"; but "we have to think again about the sources of that which is not corporate; of those practices, experiences, meanings, values which

are not part of the effective dominant culture. We can express this in two ways. There is clearly something that we can call alternative to the effective dominant culture, and there is something else that we can call oppositional, in a true sense."

11. In this connection, see John F. Kasson, *Civilizing the Machine: Technics and Republican Values in America, 1776–1900* (New York: Grossman, 1976); Thomas Reed West, *Flesh of Steel: Literature and the Machine in American Culture* (Nashville, Tenn.: Vanderbilt University Press, 1967); and Carroll W. Pursell, ed., *Readings in Technics and American Life* (New York: Oxford University Press, 1969).

12. See the invaluable collection of data compiled and organized by Rosa María Brenca de Russovich and María Luisa Lacroix, "Radio y poder en la Argentina, 1920–1953 (Notas para una historia de la radio en la Argentina)," *Cuaderno de Periodismo* 1, n.p. [*Cuaderno de Periodismo* is a publication of the Facultad de Ciencias Sociales, Universidad de Lomas de Zamora.—Trans.]

Chapter 1

1. José M. Delgado and Alberto J. Brignole, *Vida y obra de Horacio Quiroga* (Montevideo: La Bolsa de los Libros, Biblioteca Rodó, 1939).

2. Ibid., 56–57.

3. Ibid., 58.

4. To judge from the records of inventions in the office of trademarks and patents, electrolytic galvanization was beginning to be of interest to foreign enterprises based in Buenos Aires as well as to creative citizens of the River Plate Basin, who took to it with almost the same level of excitement they had brought to advances involving photography and the phonograph. See "Electroquímica, electrólisis, galvanoplastia y similares; medios, aparatos, etc., empleados y composición o preparación de baños electrolíticos; electrodos y electrolitos; sus diferentes aplicaciones en la industria," Registro de Marcas y Patentes, group 2, drawer 26. Before 1913, five undated inventions were patented, three of them Argentine and two of them renewals of foreign patents. In photography, also before 1913, sixteen patents were granted, of which only two were renewals of North American inventions. But the truly spectacular increase, in all areas, came toward the beginning of the 1920s, in 1923, when the number of annual patents jumped from around 1,000, in the years between 1910 and 1919, to more than 2,800.

5. See Delgado and Brignole, *Vida y obra de Horacio Quiroga*, 299: "He had acquired a great store of knowledge about industrial physics and chemistry, as he had about everything related to the trades. The Hoepli

technical manuals were his favorite reading matter." [Hoepli, an Italian publishing firm founded in 1870, is known for its handbooks and manuals.—Trans.]

6. Horacio Quiroga, "Los destiladores de naranja," in Napoleón Baccino Ponce de León and Jorge Lafforgue, eds., *Todos los cuentos: Horacio Quiroga* (Nanterre/Madrid: Allca XX/Fondo de Cultura Económica, 1993); Delgado and Brignole, *Vida y obra de Horacio Quiroga,* 224.

7. In this respect, Quiroga's fiction is different from the fiction that other modernists based on various "scientific" hypotheses. To take the inevitable example, nothing could be farther from Lugones than these pragmatic concerns, altogether foreign to the tone of the stories collected in Leopoldo Lugones, *Las fuerzas extrañas*, ed. Pedro Luis Barcia (Buenos Aires: Ediciones del 80, 1987, orig. 1906).

8. See Horacio Quiroga, "Cadáveres frescos," in *Obras inéditas y desconocidas*, ed. Angel Rama (Montevideo: Arca, 1968), 130ff.

9. Delgado and Brignole, *Vida y obra de Horacio Quiroga,* 300–301 (emphasis added): "In the evenings, when he left the Office of the Consul, he would join a group of friends at El Toyo café, on Calle Corrientes between Reconquista and San Martín. Later he would leave them and go on alone to the Ferretería Francesa, which he visited almost daily. He spent hours at a time there examining equipment and tools, or trying to get this or that type of screw, or dyes, or chemical substances. . . . Otherwise he would head for the bookshops and wear himself out tracking down the latest news, especially by riffling through the manual-trade digests, *which attracted him more powerfully than any other kind of book,* and of which he ended up having a remarkably complete collection." [On Beatriz Sarlo's reference to "the zeal of a Benjaminian collector," see Walter Benjamin, *Illuminations*, ed. Hannah Arendt, trans. Harry Zohn (New York: Schocken Books), 1969, and Walter Benjamin, "The Work of Art in the Age of Mechanical Reproduction," in David Richter, ed., *The Critical Tradition: Classic Text and Contemporary Trends* (New York: St. Martin's Press, 1989).—Trans.]

10. Roberto Arlt, "El paraíso de los inventores," *El Mundo,* Jan. 28, 1931. See also chapter 4, "Inventors: Technology and Mythmaking."

11. Unlike Jorge Newbery, another dandy of the period, Quiroga did not become an aviator, but he did give flying a try, executing loops and other aerobatic stunts.

12. [The Italian phrase refers to the famous English company Liberty Ltd, one of the few to have been associated with the art nouveau movement; see also chapter 2, n. 10, regarding Alejandro Sirio.—Trans.]

13. [Filippo Tommaso Marinetti (1876-1944), Italian poet, editor, founder and chief spokesperson of the international art movement

known as Futurism, and author of a series of declarations called the *Futurist Manifestos,* the first of which appeared in 1909; Gabriele D'Annunzio (1863–1938), Italian poet, novelist, dramatist, and soldier; Oliverio Girondo (1891–1967), Argentine poet, and author of *Espantapájaros* and other works.—Trans.]

14. [Dorothy Phillips (1889–1980) appeared in several early Lon Chaney films, including *Fires of Rebellion* (1917), *Hell Morgan's Girl* (1917), *Pay Me* (1917), and *Paid In Advance* (1919). The last film in which she appeared was *The Man Who Shot Liberty Valance* (1962).—Trans.]

15. [The Lumière brothers, Auguste (1862–1954) and Louis (1864–1948), French photographic manufacturers, inventors, and pioneering filmmakers, who in 1895 invented an early motion-picture camera that also functioned as a projector. The original French titles of their films *Feeding the Baby* and *Watering the Gardener* were, respectively, *Les repas de Bébé* and *L'Arroseur arrosé.*—Trans.]

16. [Marie-Georges-Jean Méliès (1861–1938), an early pioneer of French filmmaking, whose best-known motion picture is *A Trip to the Moon* (1902).—Trans.]

17. Carlos Dámaso Martínez has studied this aspect of Quiroga's work, and his conclusions have facilitated my approach to the subject. See Carlos Dámaso Martínez, "Horacio Quiroga: La fascinación del cine y lo fantástico" in the "Cultura y Nación" issue of *Clarín*, Buenos Aires, Mar. 5, 1987, and Carlos Dámaso Martínez, "Horacio Quiroga: La búsqueda de una escritura," in David Viñas and Graciela Montaldo, eds., *Historia social de la literatura argentina*, vol. 7: *Yrigoyen entre Borges y Arlt (1916–1930)* (Buenos Aires: Contrapunto, 1989). On Quiroga's story "El vampiro," see also Annie Boule, "Science et fiction dans les contes de Horacio Quiroga," *Bulletin Hispanique* 72 (1970), 3–4. In addition to short stories, Quiroga published a number of articles about motion pictures in *Caras y Caretas* (1919 and 1920, under the pseudonym "El esposo de Dorothy Phillips," an allusion to one of his characters); in *Atlántida* (1922); and in *El Hogar* (1927).

18. Horacio Quiroga, "El retrato," in *El crimen del otro* (Buenos Aires: Editorial Spinelli, 1904).

19. [See Aldofo Bioy-Casares, *La invención de Morel* (Buenos Aires: Editorial Losada, 1940), a novel depicting an island where recordings of events enable the dead to relive them repeatedly.—Trans.]

20. For records on a variety of local inventors between 1916 and 1922, see the Registro de Marcas y Patentes, group 2, drawer 26. In 1922, the number of patents—local as well as renewed foreign patents, and patents related to still photography as well as to film—had reached more than thirty, whereas before there had been only a few per year. Notable among

them is number 18.175, the patent for a screen invented by Lola Mora, which was to be used for watching movies in daylight.

21. Quiroga, "Los destiladores de naranja." ["Still very much with them in October": recall that in the southern hemisphere, spring arrives in September.—Trans.]

22. Horacio Quiroga, "Los fabricantes de carbón," in Ponce de León and Lafforgue, eds., *Todos los cuentos.*

23. Ibid.

24. [Eduardo Ladislao Holmberg (1852–1937), German-born Argentine naturalist and writer, and longtime director of the botanical gardens in Buenos Aires. His works include tales of the fantastic as well as many treatises on botany, zoology, and biology.—Trans.]

25. In this connection, see Hugo Vezzetti, *La locura en la Argentina* (Buenos Aires: Folios, 1983), and Hugo Vezzetti, *El nacimiento de la psicología en la Argentina* (Buenos Aires: Puntosur, 1988).

26. For a study of the meetings of the Sociedad Luz de Buenos Aires, see Dora Barrancos, "Ciencia y trabajadores. La vulgarización de las tesis darwinianas entre 1890 and 1920," paper presented to the Jornadas Inter Escuelas de Historia, Facultad de Filosofía y Letras, Universidad de Buenos Aires, Sept. 1991. [See also Dora Barrancos, *Educación, cultura y trabajadores (1890–1930)* (Buenos Aires: Centro Editor de América Latina, 1991).—Trans.]

27. Horacio Quiroga, "Idilio," in *El crimen del otro.*

28. Horacio Quiroga, "Una historia inmoral," in *Cuentos,* ed. Jorge Ruffinelli, vol. 4 (Montevideo: Arca, 1968). The story was first published in *Nosotros* 1:5 (1907).

29. Horacio Quiroga, "Los perseguidos," in *Los Perseguidos* (Buenos Aires: Arnaldo Moen, 1905).

30. [Elías Castelnuovo (1893–1980), typographer, linotypist, novelist, member of the literary Generation of 1920, and principal force behind the Boedo group of working-class writers, whose sociopolitical engagement and plebeian themes set them apart from the Florida group, which included Oliverio Girondo and Jorge Luis Borges. Among Castelnuovo's works, which reflect his interest in medicine and feature the ravages of "social disorders," are *Tinieblas* and *La Editorial.*—Trans.]

31. Horacio Quiroga, "El haschich," in *El crimen del otro.*

32. See, for example, Horacio Quiroga, "Una estación de amor," in *Cuentos de amor, de locura y de muerte* (Buenos Aires: San Ignacio Sociedad Cooperativa, 1917), and Horacio Quiroga, "El infierno artificial," *Fray Mocho* (1913).

33. In this connection, see also Horacio Quiroga, "El perro rabioso," the tale of an episode of hydrophobia-induced madness, in Ponce de León and Lafforgue, eds., *Todos los cuentos.*

34. See the well-informed prefatory study by Pedro Luis Barcia in Lugones, *Las fuerzas extrañas*, which details the extraliterary sources of Lugones's tales and the origin, in one of them ("Yzur"), of Quiroga's "El mono ahorcado."

35. See Barrancos, "Ciencia y trabajadores," 7: "The success of Haeckel's work in our country, and also, I believe, in Spain, where the first translations were done, was so pronounced that I am inclined to think that more than a few people were introduced to Darwin's ideas through reading texts by Haeckel." In the paper cited here, this opinion is based on texts and lectures that were published and presented in Buenos Aires.

36. Horacio Quiroga, "Historia de Estilicón," in *El crimen del otro*. Quiroga also published another story about monkeys; see Horacio Quiroga (writing as S. Fragoso Lima), *El mono que asesinó* (serial novel), *Caras y Caretas* 552–57 (1909). *El mono que asesinó* and *El hombre artificial*, along with other serial novels that Quiroga published in *Caras y Caretas* under this pseudonym, were reprinted in Horacio Quiroga, *Novelas cortas* (Havana: Arte y Literatura, 1973), with a concluding study by Noé Jitrik in which these stories are said to "form part of the wave of fantastic literature that found its supreme expression in 1906 with *Las fuerzas extrañas*, by Leopoldo Lugones."

37. Horacio Quiroga (writing as S. Fragoso Lima), *El hombre artificial*. Like Lugones's "La fuerza Omega," *El hombre artificial* is a work of fiction that is scientific in nature. For the differences between the two tales, it should be possible to study the movement from a fantastic narrative that contains "scientific" material (Lugones) to a text, like Quiroga's, that belongs more clearly to the territory of science fiction, which had yet to be developed in North American and European literature. On the primitive science fiction that originated in North America and Europe, see Sam Moskowitz, *Science Fiction by Gaslight: A History and Anthology of Science Fiction in the Popular Magazines* (Westport, Conn.: Hyperion Press, 1968).

38. [The year 1908 may be an allustion to the date when the Nobel Prize in chemistry was awarded to Ernest Rutherford (1871–1937), who was credited with, among other things, "investigations into the disintegration of the elements."—Trans.]

39. Since the end of the eighteenth century, the notion that strong feelings can be produced and transmitted by electrical or other kinds of waves had been present in essays about mesmerism; see Robert Darnton, *Mesmerism and the End of Enlightenment in France* (Cambridge, Mass: Harvard University Press, 1968). One curious fact deserves mention: a primitive science fiction story, "Dr. Materialismus," in which strong feelings are induced by electrical charges, was written by Frederic Jessup Stimson, U.S. ambassador to Argentina at the beginning of the twentieth century.

Chapter 2

I am indebted to Adrián Gorelik and Graciela Silvestri for the epigraph to this chapter, and for the discussion of several ideas that appear in the chapter's first section.

1. Where the construction of a vision is concerned, important viewpoints are presented by Paul Virilio, *La machine de vision* (Paris: Galilée, 1988).

2. In his famous 1929 visit to Buenos Aires, Le Corbusier referred to the urban history of Buenos Aires more as a metaphor than as a problem that needed to be addressed by city planning. His metaphor evoked the fortifications of the colonial city, and its unserious treatment in the famous lectures of 1929 clearly shows that Le Corbusier did not think there were any urban-historical problems to be solved in Buenos Aires (as there clearly were, for him, in older European cities). But this reference to colonial fortifications allowed him to reiterate the importance of the river, and to present the members of his local audience (who were, for their part, very concerned about history) with a decent image of the past. On this visit by Le Corbusier and his projects for Buenos Aires, see Pancho Liernur and Pablo Pschepiurca, "Precisiones sobre los proyectos de Le Corbusier en la Argentina 1929/1949," *Summa* (Dec. 1987). For parallels between Le Corbusier and Arlt, see Beatriz Sarlo, "Arlt: Ciudad real, ciudad imaginaria, ciudad reformada," *Punto de Vista* 42 (1992). The videotape *Buenos Aires II*, written by Adrián Gorelik, Graciela Silvestri, and Beatriz Sarlo, and directed by Rafael Filippelli, lays out Arlt's utopian urban themes.

3. [Arlt, born in Buenos Aires to immigrant parents, was a first-generation Argentine.—Trans.]

4. On this point, Arlt differs not only from Jorge Luis Borges but also from the Creole elites, from the most traditional nationalist factions as well as from the vanguardist or modernizing groups. On the influence of national mythologies in the formulation of utopias and projects for transformation, see the excellent study by Bronislaw Baczko, "Utopia," in *Los imaginarios sociales: Memorias y esperanzas colectivas* (Buenos Aires: Nueva Visión, 1991). On the intellectual establishment and cultural ideologies of this period, see María Teresa Gramuglio, "El campo literario en la década del 30: Imágenes de escritor, proyectos literarios y espacios de legitimación," unpublished paper, CIUNR (Consejo de Investigaciones de la Universidad Nacional de Rosario), 1991. On portrayals of the city and the sense of history, see Adrián Gorelik and Graciela Silvestri, "El pasado como futuro: Una utopia reactiva en Buenos Aires," *Punto de Vista* 42 (1992).

5. Roberto Arlt, "Noche terrible," in *Obras completas*, vol. 1 (Buenos Aires: Carlos Lohlé, 1981), 819. [All subsequent references in this chapter to Arlt, *Obras completas*, are also to vol. 1.—Trans.]

6. Roberto Arlt, *Los lanzallamas*, in *Obras completas*, 414.

7. [Antonio Berni (1905–1981), Argentine painter and designer of the Platine School.—Trans.]

8. See Arlt, *Los lanzallamas*, 349ff, for the Melancholy Ruffian's route from the corner of Maipú and Diagonal to Corrientes.

9. Roberto Arlt, *Los siete locos*, in *Obras completas*, 129: "His sorrow, completely decompressed, reached all the way to the horizon, glimpsed through trolley and streetcar cables." The same sky was described by a writer for *Crítica* (Aug. 20, 1920): "It's possible that not too many people have noticed an item as simple as it is wonderful. Hidden among chimneys, air ducts, telephone poles, and electric wires, and scattered along the roofs of houses in this sensitive, alert city, are quite a few radio antennas." On the reasons for the mass-circulation newspapers' ideas about technology, see chapter 3, "Popular Science and the Popularizing Press."

10. [Alejandro Sirio, born Nicanor Alvarez Díaz in 1890 in Oviedo, Spain, was one of the most celebrated illustrators and caricaturists of his day. He moved to Argentina in 1908 and worked first as an illustrator for the Sastrería Inglesa (see also chapter 1, n. 12, regarding the *stile Liberty*) and later at *Caras y Caretas*, *Plus Ultra*, *La Nación*, and other publications. One of his notable caricatures, produced in 1938 to accompany a newspaper profile, was of the writer Leopoldo Lugones. Sirio died in Buenos Aires in 1953.—Trans.]

11. Roberto Arlt, "Corrientes por la noche," in Daniel Scroggins, ed., *Las aguafuertes de Roberto Arlt* (Buenos Aires: ECA, 1981), 147–48.

12. Roberto Arlt, *El amor brujo*, in *Obras completas*, 611: "Balder smiles faintly. He watches the mass of skyscrapers, pink and gray, leaning into a night so dense that the darkness could be a hill pierced by shining holes. The sudden dimming of a white-and-violet neon sign. Whistles of far-off locomotives. Men walking by in caps and tunics. 'Yes, it's nice, the city.' "

13. [Remo Erdosain is the protagonist of *Los lanzallamas* and *Siete locos*.—Trans.]

14. Arlt, *Los lanzallamas*, 471–72.

15. Wladimiro Acosta worked on a series of studies for his integral city block between 1927 and 1935. On Acosta, see especially Pancho Liernur, "Wladimiro Acosta y el expresionismo alemán," and Anahi Ballent, "Acosta en la ciudad: Del city-block a Figueroa Alcorta," in *Wladimiro Acosta, 1900–1967* (Buenos Aires: Facultad de Arquitectura y Urbanismo,

Universidad de Buenos Aires, 1987), which served as the catalog for an exhibition of the same name, held in Buenos Aires.

16. Wladimiro Acosta, "Serie de estudios sobre 'city-block,'" in *Vivienda y ciudad: Problemas de arquitectura contemporánea* (Buenos Aires: Anaconda, 1947), 140.

17. See Acosta, *Vivienda y ciudad*, especially the introduction and 134–35, where Acosta lays out the need for the integral city block.

18. Arlt, *El amor brujo*, 566.

19. See Ballent, "Acosta en la ciudad," 33, where the author astutely points out the external nature of Acosta's intervention as well as the gap between the urban reality of Buenos Aires and the different versions of the integral city block: "These were buildings that did not blend into the existing urban fabric. The very development of the integral city block was symptomatic; even the first proposal for a cruciform building clearly showed its discontinuity with the city's development. But that version preserved basic dimensions that might have enabled its assimilation into the blocks of buildings in Buenos Aires; it did not change the structure of the weft. By contrast, the integral city block was based on the recognition that the goals of the new plans could not be reconciled with attachment to the existing grid." [The term "urban renewal" is something of an anachronism in that it dates from 1954; in this and other contexts, however, it seems more idiomatic and more accurate than either "urban reform" (*reforma urbana*) or "reformism" (*reformismo*).—Trans.]

20. See Liernur, "Wladimiro Acosta y el expresionismo alemán," 19: "The premise set forth in this essay consists in the understanding that the quest for 'the South' did indeed constitute the chief motive of Acosta's work and trajectory, at least until 1940, and that, over the course of a 'long expedition,' as a personal motive it gave shape to a geographically extreme episode in the literal pursuit of the Sun and the Light that defined the faction of the artistic vanguard with which he had been linked in the German years."

21. See Acosta, *Vivienda y ciudad*, 152, where it is explained that integral city blocks were divided into two types, A and B, and that those of type A had "the following fundamental characteristics: (1) Juxtaposition of living and work space. (2) Elective uniform orientation of both, with maximum exposure to the sun for residences, and uniform diffuse lighting for workplaces. (3) Continuity of green space, and development of large parks. (4) Differentiation of traffic, and elimination of perfectly level intersecting lines. In each integral city block, the structure consists of a large, linear block of residences, 100 meters high, facing north, and four office buildings, perpendicular to the residential block and lower in height

. . . . The ground and first floors are occupied by businesses. A longitudinal access street runs parallel to the residential block, passing underneath the office buildings, which form bridges over it. The first-floor businesses themselves pass over the transverse access streets. In this way, a network of internal low-level streets for pedestrians is created within each integral city block." Integral city blocks of type B were an improved version of type A, with the following features; see ibid., 156: "(1) Completely differentiated traffic. Separation not just between access streets and fast-transport streets but also between roadways for motor vehicles, at the upper level, and for pedestrians, at the ground-floor level. The ground belongs entirely to pedestrians. (2) Absolute continuity of protected roadways for pedestrians. (3) Shortening and elevation of the office buildings, with resulting reduction in the ground surface they occupy, in favor of open space, and with no change in their capacity. (4) Development of large sites with zenithal lighting, for factories, workshops, concert halls, lecture halls, cinemas, theaters, etc." ["The first-floor businesses": recall that in South America, as in Europe, a distinction is made between the ground floor and the first floor, the latter corresponding to what would be called the second floor in the United States.—Trans.]

22. Ibid., 131.

23. Arlt, *Los lanzallamas*, 325.

24. Ibid., 328.

25. See Roberto Arlt, "La luna roja," in *Obras completas*, 751ff, a text specifically worked out through images from movies and movie posters. The dance hall, where news of the impending catastrophe is first heard, comes directly from cinematic visual experience; nevertheless, the city of the tale is a science fictional setting in which all the elements of architectural and technological modernity—high-voltage wires, overlapping fluorescent facades, electrical signs, aerial gardens, and so on—are intensified.

26. Arlt, *El amor brujo*, 566.

27. Roberto Arlt, "Aventura sin novela y novela sin aventura," *El Mundo*, Aug. 13, 1941.

28. Roberto Arlt, *Obras completas*, 741, 819, 321, 352. This stylistic feature has already been pointed out by Noé Jitrik, "Entre el dinero y el ser," *Dispositio* 1:2 (1976). See also Beatriz Sarlo, *Una modernidad periférica: Buenos Aires 1920 y 1930* (Buenos Aires: Nueva Visión, 1988), 58.

29. Arlt, *Los siete locos*, 179.

30. [Allan Kardec was a pseudonym of Hippolyte Léon Denizard Rivail (1804–1869), who systematized Spiritism.—Trans.]

31. Arlt, *Los siete locos*, 295.

32. Ibid., 211–12.

33. George Orwell, Yevgeni Zamiatin, and Aldous Huxley, among others, plotted allegorical political fiction that was based on the power of science in the service of what was, in no uncertain terms, a totalitarian order.

34. [Invention of Sir William Crookes (1832-1919), British chemist and physicist, whose most important investigations involved the conduction of electricity in gases; the Crookes tube is where cathode rays were first produced.—Trans.]

35. Arlt, *Los lanzallamas*, 473.

36. Ibid., 453–55.

37. The Astrologer's monologue at the end of *Los siete locos* includes a toxic-gas factory, automobiles plated with iron-and-nickel chrome, bands of assassins in airplanes, cells in radiotelegraphic communication, hydroelectric projects, chemical and biological weapons, cinematography as an instrument of propaganda, industrial landscapes with blast furnaces and clouds of fire, cannon factories, insurrections set off by radio signals, Bessemer converters, germ warfare, and tactics centered on the control of long-distance communications. This technological apparatus is intended to enable a revolution directed by an absolute minority of utterly fearless leaders. Erdosain's final reverie, also in *Los siete locos*, is composed of cinematic projections onto the clouds and the invention of the death ray (a term also used by newspapers of the period to refer to a number of new types of weapons). It is not particularly surprising that Erdosain should emerge from his depression with fantasies of transforming himself into the "king of the universe"; what is surprising is the technological material of this transformation. The technological details of the Arltian narrative share a feature outlined by the authorial preface to *The Scientific Romances of H. G. Wells* (London: Gollancz, 1933), where Wells observes that, except in tales of exploration, fantastic elements in fiction traditionally had depended on the intercession of magic (with even Dr. Frankenstein's man-made monster needing a touch of magic in order to come to life), but that it became impossible toward the end of the nineteenth century to elicit even passing belief in magical phenomena, whereupon Wells, understanding that he could substitute scientific language for traditional encounters with magicians or the devil, updated his material and approached it in terms of actual scientific theories.

38. [Georges Sorel (1847-1922), French social philosopher and journalist, was the leader and principal theoretician of the revolutionary syndicalist movement.—Trans.]

39. In this regard, see two careful studies: José Amícola, *Astrología y fascismo en la obra de Roberto Arlt* (Buenos Aires: Weimar, 1984), and Jorge

Rivera, *Los siete locos* (Buenos Aires: Biblioteca Crítica Hachette, 1986). An intelligent interpretation of the relationship between literature and machinery is presented by Alan Pauls, "Arlt: La máquina literaria," in David Viñas and Graciela Montaldo, eds., *Historia social de la literatura argentina*, vol 7: *Yrigoyen entre Borges y Arlt (1916–1930)* (Buenos Aires: Contrapunto, 1989).

40. On the various types of premises in science fiction, see Samuel Delany, "About Five Thousand One Hundred and Seventy-Five Words," in Thomas Clareson, ed., *Science Fiction: The Other Side of Realism* (Bowling Green, Ohio: Bowling Green University Popular Press, 1971). On the relationship between science and science fiction, see Joanna Russ, "Towards an Aesthetic of Science Fiction," in R. D. Mullen and Darko Suvin, eds., *Science-Fiction Studies: Selected Articles on Science Fiction, 1973–1975* (Boston: Gregg Press, 1976), 13, where Russ describes science fiction as "the only modern literature which attempts to assimilate imaginatively scientific knowledge about reality and the scientific method." See also David Ketterer, *New Worlds for Old: The Apocalyptic Imagination, Science Fiction and American Literature* (Garden City, N.Y.: Anchor Books, 1974), 24: "In general terms, the proliferation of science fiction is a response to abruptly changing social conditions. During times of stability, when change neither happens nor is expected, or happens so gradually as to be barely noticeable, writers are unlikely to spend time describing the future conditions of society, because there is no reason to expect any significant difference. With the nineteenth century, things speeded up, and now change is a constant and unnerving factor in our daily lives. If we are to live rationally, and not just for the moment, some attempt must be made to anticipate future situations." Finally, see Gary Wolfe, *The Known and the Unknown: The Iconography of Science Fiction* (Kent, Ohio: Kent State University Press, 1979), 4–5: "Science fiction . . . provides technological society with a ritual methodology for action. In its simplest formulation, this methodology is the scientific method of systematic induction. . . . Science fiction . . . explore[s] the mythical aspects of reason itself, specifically of scientific reasoning."

41. To Eduardo Fernández, current president of the Asociación Argentina de Inventores, I owe this document as well as a long and generous conversation in which he shared his knowledge about the history of invention in Argentina.

42. See chapter 4, "Inventors: Technology and Mythmaking."

43. Roberto Arlt, *El juguete rabioso*, in *Obras completas*.

44. Arlt, *Los siete locos*, 256–57.

45. Ibid., 257.

46. Silvio Astier, too, invents precisely those machines that were the

order of the day in newspapers like *Crítica* and *El Mundo*. To a question about what he has invented, he answers, "An automatic star-plotter, and a typewriter that takes dictation. Here's a letter of congratulations from Ricaldoni, the physicist"; see Arlt, *El jueguete rabioso*, 67. [Julio Ricaldoni (1906–1994) was an Uruguayan civil engineer and researcher affiliated with the Resilience Laboratory at the Universidad de la República Oriental del Uruguay. The laboratory, founded in 1932, was placed under Ricaldoni's direction in 1933. His first activities there involved photoresilience, a concept then new to Uruguay, but whose development was already well under way in Europe.—Trans.] This dialogue occurs in Arlt, *El amor brujo*, 659: "Victor speaks to Balder: 'Now, you, you're an engineer—do you know anything about radio?' 'No, not much . . . what interests me is architecture. . . .' 'Tell me, what do you think of those experiments they've done to blow up a mine with ultraviolet rays?' 'So far, they're inconclusive. . . .' 'But it's possible?' 'Only in a straight line, with shortwaves, within a hundred miles . . . that's all. Anything else is pure fantasy.'"

Chapter 3

1. See Sylvia Saitta, "*Crítica* en los años 30: Entre la conspiración y el exilio," *Entrepasados* 2 (1992).

2. *Crítica*, Mar. 14, 19, 20, 21, 22, 24, and 25, 1925, Apr. 20 and 24, 1925, Feb. 3, 1929, and Aug. 3, 1929; *El Mundo*, Jan. 12, 1929, Nov. 1, 1929, and Apr. 18, 1930.

3. *Crítica* published its "Radiodictionary" in 1923; Aug. 26, 1925, saw the reappearance of its "Voice of the Air" section, which had disappeared the year before, and which continued until 1927, when, again in August, it was renamed "The Weekly Radio Page." In November 1928, *El Mundo* began publication of a radio section, inspired by the success of a previous section devoted to engines and automotive technology; in March 1930, this section was renamed "Radiotelephony"; a section called "Film and Photography Industry" first appeared in January 1930. Both papers published sections on popular science that covered various other topics as well; they also published news that appeared in the body of the paper, apart from these sections.

4. Because it was not possible to consult every volume of *Crítica*, the figures that follow do not reflect the total number of articles published, even though they do reflect, very eloquently, the growth in the number of such articles, which increased steadily from the middle of the 1920s through the end of the following decade. The same thing happened at *El Mundo*.

Year	No. of months	Articles	Year	No. of months	Articles
	Crítica		1926	12	53
1916	6	4	1927	12	84
1919	3	0	1928	6	55
1920	12	1	1929	12	132
1921	8	6	1930	8	94
1922	12	9			
1923	9	44		*El Mundo*	
1924	12	62	1928	2	8
1925	12	63	1929	8	52
			1930	12	128

5. See Sylvia Saitta, "El diario *El Mundo*," unpublished paper, Universidad de Buenos Aires, 1990.

6. This information about *Crítica*'s process of technical modernization was researched by Saitta, "El diario *El Mundo*," which includes the following citation from *Crítica*, Sept. 12, 1927: "When it comes to progress in newspaper publishing, New York has met its match in Buenos Aires: only *Crítica* and the world's best big-city papers have Hoe rotary printing machines. Our rotary presses feature an amazing innovation: automatic inking, which allows complete control over print quality while the press is running, and now not just for whole pages but for individual columns, too. A single hour of continuous printing is enough for the Hoe rotary press to blanket the city with newspapers."

7. Roberto Arlt, *Los lanzallamas*, in *Obras completas*, vol. 1 (Buenos Aires: Carlos Lohlé, 1981), 519–20.

8. [An allusion to Roberto Tálice, *Cien mil ejemplares por hora: Memorias de un redactor de Crítica, el diario de Botana* (Buenos Aires: Corregidor, 1977); see the introduction to the present volume, n. 3.—Trans.]

9. See Saitta, "El diario *El Mundo*":

> As proof of its popularity, the paper continuously gave its readers and advertisers information about its print-run figures. . . . And just in case numbers alone should fail to persuade, *Crítica* certified the circulation figures with the presence of a notary public (photographed at the foot of the Hoe press) or with photos of receipts displaying quantities of paper purchased. The figures clearly show the paper's large size and leave no doubt about its constant growth:

Date	Avg. daily circulation (copies)	Date	Avg. daily circulation (copies)
Oct. 1924	166,385	Jan. 1929	251,748
May 1926	206,720	July 1929	286,923
July 1926	217,062	Jan. 1930	238,501
Mar. 1927	218,494	Mar. 1930	266,036
Aug. 1927	263,234	July 1930	274,676
Jan. 1928	224,000		

10. See chapter 4, "Inventors: Technology and Mythmaking." The figure of the inventor as cultural hero has existed in the United States since the final decades of the nineteenth century; its narrative embodiment can be found in Hugo Gernsback, *Ralph 124C 41+: A Romance of the Year 2660* (Lincoln: University of Nebraska Press, 2000), which modifies the motif of the satanic inventor, derived in large part from the science fiction novels of H. G. Wells. On this topic, see L. Sprague de Camp and Thomas Clareson, "The Scientist," in Patricia Warrick, Martin Henry Greenberg, and Joseph Olander, eds., *Science Fiction: Contemporary Mythology* (New York: Harper and Row, 1978).

11. See chapter 6, "Doctors, Clairvoyants, and Quacks."

12. For example, one series, apparently launched by the paper in June 1925, featured four monstrosities over a period of six days. *Caras y Caretas* had already shown how effective it could be to alternate short stories, news, and articles on everyday life with information about freaks of nature.

13. This is what occurred in the cases of Fernando Asuero and, to a lesser extent, Serge Voronoff; see chapter 6, "Doctors, Clairvoyants, and Quacks."

14. Among these publications, for example, were short serial novels published in weekly installments; see Beatriz Sarlo, *El imperio de los sentimientos: Narraciones de circulación periódica en la Argentina, 1917–1927* (Buenos Aires: Catálogos, 1985).

15. *Radio Cultura, ICARM, R.C.A., Radio Revista, Ciencia Popular, Revista Ondas*, and *Guía del Inventor* did certainly all exist at the same time. Toward the end of the 1920s, however, as can be inferred from collections that are far from complete, the offerings of a number of other magazines overlapped with theirs. In that period, *Ciencia Popular* was already considered the magazine that was best at popularizing technoscientific information. [*R.C.A.* was the official publication of the Radio Club Argen-

tino. *Ciencia Popular* was the Spanish-language counterpart of the North American magazine *Popular Science*. For a full explanation of the *ICARM* acronym, see chapter 4, esp. n. 42.—Trans.]

16. From the opening page of *Ciencia Popular* 4 (Nov. 1928): "We have been conducting negotiations aimed at obtaining for *Ciencia Popular* the editorial services of Dr. Jorge A. Duclout, the well-known engineer and journalist, and these negotiations have now reached a satisfactory conclusion. The success so quickly achieved by *Ciencia Popular*, both here and abroad, requires complete dedication to the magazine, something that has become impossible for us, given the reasons already cited and the increasing effort required for the steady development of *Radio Revista*, our core enterprise."

17. The first contest sponsored by *Ciencia Popular*, despite its quaint theme ("taking odd photographs in which the subjects appear to be deformed or artfully distorted"), was clearly didactic in nature, indicating an intention to promote learning about the possibilities of the lens and the iris diaphragm, and about developing and printing photographs. Issue 7 (Feb. 1929) gives examples of ideas for future contests and suggests that readers provide others: "Our readers are asked to send us their opinions of these ideas so that we can include any helpful changes in the next issue. Readers may also send us ideas for new contests, such as 'a contest involving original ideas for using old phonograph records' or 'a contest involving helpful inventions for motorists' and other ideas in this vein."

18. This notice appeared for the first time in *Ciencia Popular* 4 (Nov. 1928) and was repeated in almost every subsequent issue.

19. See chapter 5, "Radio, Cinema, and Television: Long-Distance Communication."

20. *Amazing Stories*, the first pulp magazine in the United States devoted to science fiction, appeared in 1926 and was edited by Hugo Gernsback. It also published many articles about radio electronics.

21. In connection with science fiction, see Gary Wolfe, *The Known and the Unknown: The Iconography of Science Fiction* (Kent, Ohio: Kent State University Press, 1979), 5: "What science fiction does that is uniquely suited to a technological society is to explore the mythical aspects of reason itself, specifically of scientific reasoning."

22. On the case of Brazil during the immediately preceding period—which makes Brazil even more interesting, in terms of the gap between reality and imaginary scenarios—see Francisco Foot Hardman, "Técnica e sociedade: Maquinismo como espetáculo no Brasil préindustrial," *Remate de Males* 7 (1987).

23. It was Hugo Gernsback who created this intersection of techni-

cal journalism with the literature of prognostication. In 1908 he began publishing *Modern Electrics*, a magazine devoted to electricity and radio, and he founded a society of ham radio operators. Finally, in 1911, he wrote *Ralph 124C 41+*, his optimistic epic of the future, first serialized in *Modern Electrics*.

24. *Crítica*, Oct. 23, 1927.

25. See E. M. Forster, "The Machine Stops" (1909); Carel Čapek, *RUR* (1920); Yevgeny Zamiatin, *We* (1922); Aldous Huxley, *Brave New World* (1932).

26. [The term "Fordism," reportedly coined by Antonio Gramsci (1891–1937), refers in general to the social institutions of mass production and specifically to Henry Ford's policies of using high wages, to retain workers, and assembly-line techniques, to produce mass commodities as cheaply as possible.—Trans.]

27. With respect to weaponry, *Crítica* published an interesting series of articles (May 28, 30, and 31, and June 2 and 15, 1924) on diabolical rays and efforts to reach agreement on their elimination. In both *Crítica* and *El Mundo* there was also frequent and very detailed information about torpedoes and submarines.

28. Gary Wolfe's statements about the functions of science fiction can also be applied to technical journalism; see Wolfe, *The Known and the Unknown*, 6: "One reason for the popularity of science fiction is the felt need for a body of comprehensible speculative thought in the face of the manifold complexities and specializations of modern science."

29. See Roberto Arlt, "Luna roja," in *Obras completas*, vol. 1 (Buenos Aires: Carlos Lohlé, 1981), an example of antiwar science fiction fantasy; Arlt's *Los siete locos* and *Los lanzallamas* leave unresolved the conflict involving the poetics of the machinery of war.

Chapter 4

1. [See, among his related works, Roberto Arlt, *Aguafuertes porteñas* (Buenos Aires: Victoria, 1933), the first collection of Arlt's columns for *El Mundo*.—Trans.]

2. There are a great many sources on the charismatic figure of the inventor in North American culture, and on Edison in particular. A comprehensive survey and an up-to-date bibliography can be found in Thomas P. Hughes, *American Genesis: A Century of Invention and Technological Enthusiasm* (New York: Penguin, 1989). For a provocative literary study of Brazil, see Flora Süssekind, *Cinematograph of Words: Literature, Technique, and Mod-*

ernization in Brazil, trans. Paulo Henriques Britto (Stanford, Calif.: Stanford University Press, 1997).

3. Two examples will suffice, both from *El Mundo.* The first is dated Jan. 23, 1930: "Before long, science will have found a way to completely eliminate submarines from the ocean. . . . Scientists will soon discover a microphone so sensitive and perfect that it will allow underwater tracking of a submarine's every movement. With these microphones installed in aeroplanes, aerial squadrons will be able to pinpoint the positions of these ships and destroy them with depth charges"—this, nine years before World War II. The second example from *El Mundo* is dated Apr. 19, 1930: "Today medical authorities [in Boston] were shown and offered a new radium device equipped with a special mechanism into which a man of normal size can fit. It is believed that the use of this device may make it possible to cure a variety of serious illnesses. The patient lies down inside the device, [whose vibrations] produce a kind of fever, an elevated temperature, which, it is theorized, kills bacteria. The fever rises or falls with a simple flick of a switch." After 1920, *Crítica* published the same types of news stories, among them several that evoked North American science fictional fantasy: news about the floating city of the future (June 20, 1921), animal-to-human blood transfusions (Oct. 8, 1922), the world's tiniest receiver (Sept. 6, 1923), communications with Mars (Oct. 29, 1926), selenium allowing the blind to "hear" light (Dec. 20, 1926), a new piece of information to shore up the bridge to the hereafter (Sept. 28, 1926), life created in chemistry labs (Jan. 31, 1927), veins opened, without anesthetic, for blood purification (May 7, 1927), a savant named Ferrán snatching from science the miraculous secret of rejuvenation (Oct. 16, 1927), the manufacture of monkey-humans (Oct. 19, 1927), a dirigible 235 meters long (Apr. 26, 1928), and the Opel Corporation's building of a rocket to shoot a man into space (May 2, 1928) in addition to claims that it was now understood how to get to the moon (Feb. 3, 1928). Along with these news stories, dozens of fantastic articles appeared about Dr. Fernando Asuero's treatments of the trigeminal nerve, and about implantations of chimpanzee gonads that were carried out according to Dr. Serge Voronoff's methods; see chapter 3, "Popular Science and the Popularizing Press," and chapter 6, "Doctors, Clairvoyants, and Quacks."

4. See, respectively, *Crítica,* July 24, 1921, Aug. 6, 1922, Sept. 12, 1922, Sept. 11, 1923, Sept. 12, 1923, Sept. 16, 1923, Sept. 20, 1923, and Jan. 1, 1923. In the month of September, a large number of inventors entered a contest sponsored by the Buenos Aires Radio Club on the occasion of the International Radio Exposition of 1923, to which *Crítica* gave particularly thorough coverage.

5. To give just one example, see a very important text by Richard Maclaurin, "Mr. Edison's Service for Science," in Carroll W. Pursell, ed., *Readings in Technology and American Life* (New York: Oxford University Press, 1969).

6. *El Mundo*, Mar. 30, 1930. Another case of competition in which an Argentine patent preceded the foreign one was that of Luis Guajardo, who had invented a mechanical device, activated by a wind-up or electrical motor, for summoning help in danger; see *El Mundo*, June 4, 1930. The newspaper spells out the moral of the story: "Thus we are shown that in our country, too, there are noteworthy technicians who, in the spirit of tenacity, quietly celebrate accomplishments such as the one apparently indicated in this case."

7. Long articles were devoted to these cases in *Crítica*, Jan. 1, 1926, Jan. 29, 1926, Dec. 28, 1926, Feb. 3, 1928, Apr. 7, 1928, June 21, 1928, Feb. 7, 1929, Feb. 25, 1929, and Sept. 22, 1929. Magazines like *Ciencia Popular*, *R.C.A.*, and *Radio Revista* constantly published this type of material, especially in letters from readers. [Again, *R.C.A.* was the official publication of the Radio Club Argentino.—Trans.]

8. The reporters at *Crítica* launched a fund for Parpagnoli to travel to the United States, and they recommended him to the local Italian community, which must have felt a moral obligation to help him; see *Crítica*, Mar. 24, 1924.

9. Roberto Arlt, "El paraíso de los inventores," *El Mundo*, Jan 28, 1931.

10. The papers customarily published items about the granting of patents, a sign of how important such news was from a journalistic standpoint.

11. *Crítica* was particularly active in exposing medical or psychiatric fraud but also took care not to miss out on news about miracle cures.

12. "La cosmografía elemental al alcance de los niños: D. Antonio Scanga a inventado un aparato simple como ingenioso," *El Mundo*, Feb. 7, 1929.

13. Roberto Arlt, *El juguete rabioso*, in *Obras completas*, vol. 1 (Buenos Aires: Carlos Lohlé, 1981), 31, 33.

14. Silvio Astier, asked how he came to imagine some of his inventions, says, "I'll be walking down the street, and in some mechanic's shop I'll see a machine I'm not familiar with. I'll stop and say to myself, while studying the different parts of what I'm looking at, This must work like this or that, and it must be used for such and such"; see ibid., 69. For another perspective on this topic, see Alan Pauls, "Arlt: La máquina literaria," in David Viñas and Graciela Montaldo, eds., *Historia social de la literatura argentina*,

vol 7: *Yrigoyen entre Borges y Arlt (1916–1930)* (Buenos Aires: Contrapunto, 1989).

15. See Luis Alberto Romero, "Libros baratos y cultura de los sectores populares," unpublished paper, Centro de Investigación Sobre el Estado y la Administración (CISEA), 1986; and Leandro Gutiérrez y Luis Alberto Romero, "Sociedades barriales, bibliotecas populares y cultura de los sectores populares: Buenos Aires, 1920–1935," *Desarrollo Económico* 109 (Dec. 1989).

16. *Enseñanza por correo—Guía del Exito*, free pamphlet distributed by the Ateneo Técnico y Comercial, n.d., 55 (the pamphlet is undated, but the diploma in the illustration is dated 1933). Such pamphlets had the same characteristics throughout this period, as newspaper and magazine advertisements also did. In many cases, they included quotations and famous sayings of North American millionaires or presidents, clearly revealing some of the sources that inspired the advertising copywriters for this new form of distance education, which was also advertised in magazines like *Caras y Caretas*. The language that was used to inspire prospective students was the traditional language of self-help books, which had already existed for decades in North America.

17. Ibid., 24.

18. T. A. Le Breton, *Manual del inventor: Con las leyes de patentes, marcas, identificación y monopolios* (Buenos Aires: Le Breton–Obligado & Cía, 1928). See also Elisa Bachofen, *Guía del inventor* (Buenos Aires: Círculo Argentino de Inventores, 1927).

19. Statistics support this conviction; see Le Breton, *Manual del inventor*, 124: "Of one thousand patents granted in the United States in 1912, more than 65 percent were for inventions that had no connection with the occupations of the inventors."

20. Ibid., 134.

21. See ibid., 138–53, where "topics for inventors" take up fifteen pages and include potable-water filters; milk containers; nonrefillable bottles; electric-lamp protectors; a fire alarm and extinguisher; wax matches for windy days; instant tire changes; automatic speed changes; a vacuum for the dust churned up by automobiles; rail junctions; a device for eliminating smoke from locomotives; automatic halting of derailed trains; self-closing barricades; rail repair; a filter for noxious exhaust fumes; uses for the power of tides; durable accumulators and batteries; molds for house construction; a bricklaying device; roof repair; wall dehumidification; a wallpapering device; uses for sawdust and other waste products of sawmills; indestructible street repairs; the transformation of latex waste products; the manufacture of artificial latex; an oil-purification device; a broom-stitching machine;

a rideable motorized plow; a potato reaper; a mechanical hoe; a multipurpose gardening tool; a device for picking fruit, peanuts, cotton, and sugarcane; corn huskers; portable corrals; automatic feeders for corrals; an automatic fowl-fattening device; a machine for sealing envelopes; and so on, and so forth.

22. I am indebted to Eduardo Fernándea, president of the Asociación Argentina de Inventores, for information about the economic fate of independent local inventors before and after the arrival of Biro. On Biro himself, see E. Fernández, "Ladislao José Biro, el inventor de la birome," *Boletín de la Asociación Argentina de Inventores* 8 (1991).

23. Information provided by the review *Marcas y Patentes* and extracted from *Patentes de Invención, marcas de fábrica, de comercio y de agricultura*, official publication of the Office of Patents and Inventions, shows the growth in patents as follows, although it does not cover all the years between 1890 and the 1930s:

Year	No. of patents	Year	No. of patents	Year	No. of patents
1890	142	1901	231	1912–22	no data
1891	141	1902	243	1923	2,893
1892	125	1903	282	1924	1,538
1893	118	1904	275	1925	1,586
1894	141	1905	368	1926	no data
1895	181	1906	593	1927	no data
1896	159	1907	631	1928	2,106
1897	256	1908	543	1929	1,948
1898	262	1909	780	1930	1,623
1899	216	1910	876	1931	925
1900	264	1911	995	1932	3,555

The growth in patents was remarkable, even though these figures cannot be compared to those for the United States (where more than 40,000 patents were granted in 1924; in Germany, as in France, there were 18,000; in Italy, 6,468).

24. See Luis de Lemoine, *Inventos y marcas* (Buenos Aires: Estudio Lemoine, Inventos y Marcas Editores, 1943), where these figures are broken down by nationality: 1,269 inventors from Argentina, 582 from the United States, 297 from England, 195 from Germany, and 133 from France.

25. ["Have you made an interesting discovery? . . ." See chapter 3, esp. n. 18.—Trans.]

26. *Ciencia Popular* 12 (July 1929).

27. For these examples, see *Ciencia Popular* 8 (Mar. 1929). In the years that followed, the "Practical Ideas from Readers" page appeared with great frequency.

28. For example, see "Morse y su invento," *Ciencia Popular* 14 (Sept. 1929), with the significant subtitle "Hay quien dice que el genio es un desbordamiento de la razón, quien lo define como la esencia de la humanidad, y quien lo denomina el espíritu de los siglos" (There are those who say that genius is a superabundance of reason, those who define it as the essence of humanity, and those who label it the spirit of the ages).

29. In the organization's memorandum of 1923, members of the following institutions figure as temporary members (a category designating those invited to visit the Círculo for a particular length of time or to carry out some activity at its headquarters): the Centro Nacional de Ingenieros, the Círculo Militar, the Sociedad Argentina de Estudios Geográficos, the Sociedad Científica Argentina, the Touring Club Argentino, the Aero Club Argentino, the Club del Progreso, the Sociedad Rural Argentina, the Círculo Médico Argentino, the Unión Popular Católica Argentina, the Jockey Club, the Círculo Tradición Argentino, the Liga Patriótica Argentina, the Unión Industrial, and the Centro Naval Militar.

30. See *La Nación*, Dec. 9, 1922; also reprinted in *ICARM* 1:1–9 (Jan.–Sept. 1923), the first issue of the Círculo's magazine.

31. [Thirty-three, in Spanish, is *treinta y tres*, and Treinta y Tres is the name of Uruguayan province number 19 as well as of the province's capital; the namesakes of both are thirty-three patriots who are honored by Uruguayans.—Trans.]

32. Among thirteen who did not attend the meeting but conveyed their support were seven academics, four of whom were engineers. Two years later, in notices that appeared in the Círculo's magazine, Carlos J. Rodríguez, a national representative who had sponsored a bill for a law to protect inventors, was mentioned as an active member; see *ICARM* 4:1–12 (Jan.–Dec. 1926). In 1929, the steering committee had twenty-four members and included fifteen academics and three military officers.

33. The Army Mechanics School was cited a number of times in publications read by amateur inventors and technology enthusiasts; see, for example, *Ciencia Popular* 4 (Nov. 1928). The school, founded in 1902 by General Pablo Ricchieri, started out with sixty enrollees; by the middle of the 1920s, there were already more than four hundred. The faculty included many more civilians than military personnel, as indicated by a photograph from the period, showing sixteen civilians and six military officers.

34. Bylaws, article 55, clause (a).

35. Bylaws, article 60.

36. Bylaws, article 11.

37. This was so clearly one of the Círculo's purposes that the *Guía periodística argentina* (Buenos Aires: Lerose y Montmasson, 1925) describes the organization's magazine as a publication that "offers investors the latest interesting and practical Argentine inventions, putting inventors in touch with investors and industrialists, not pursuing profits as a goal, and protecting, in the broadest sense, any inventor who has an Argentine patent." In the Círculo's analytical report on its first eight years, table 1 is titled "Official Actions Taken on Behalf of Members" and deals almost exclusively with the promotion of inventions to public organizations or private enterprises; see Círculo Argentino de Inventores, *Memoria sintética de los ocho años de vida del Círculo Argentino de Inventores* (Buenos Aires: Círculo Argentino de Inventores, 1930). According to that report, 223 inventions merited this kind of attention from the Círculo between 1922 and 1930, although the organization's magazine publicized many more. In the same report, table 9 features the names of eighty-four inventors whose inventions were offered to industry. The Círculo's dedication in performing this task was also reflected in efforts undertaken with official embassies in Argentina. In 1923, for example, Ruiz Moreno and Elisa Bachofen approached the Japanese ambassador, on the assumption that Japan was a country that might be interested in a "new, patented system for building low-cost, earthquake-proof houses." The ambassador politely replied that he had forwarded the offer to his government; see *ICARM* 1:10 (Oct. 1923). Naturally, nothing ever came of it.

38. See Elisa B. Bachofen, "Cómo patentar un invento," *ICARM* 4:1–12 (Jan.–Dec. 1926). As an example of the material difficulties, the application had to be accompanied by the following materials: "(a) Description of the invention, including an explanation of the illustrations; (b) manner in which the invention will be put into practice, that is, its construction and preprocessing; (c) replevin, that is, the basis of the right of precedence; and (d) drawings or illustrations of the invention, which must contain key letters that match the descriptions and the various views offered by the drawings. Drawings will be executed in black (india) ink, to metric scale, and in general must present objects in perspective. Further, those parts that cannot be presented in perspective must be shown in general or partial section. Two copies of these diagrams will be presented, one on onionskin and the other on Canson silk or a similar paper." See also Hugo Nemo Pantolini, "A los inventores en ciernes," *ICARM* 6:4–6 (Apr.–June 1928), an article by a mechanical engineer with the army, which does not give such detailed instructions but instead offers various examples of inventions that can only

be seen as unworkable; their creators are advised to approach the Círculo, where they will be able to get the instruction they so obviously lack.

39. Memorandum from the steering committee, presented to the regular meeting held on November 16, 1923; see *ICARM* 2:1–3 (Jan.–Mar. 1924).

40. Círculo Argentino de Inventores, *Memoria sintética de los ocho años de vida del Círculo Argentino de Inventores.*

41. See *ICARM* 5:4–6 (Apr.–June 1927). No one saw any reason why Ruiz Moreno's participation in the Liga's functions should be kept secret, and it is mentioned on various occasions in the Círculo's magazine, starting with its earliest issues. In the *Memoria sintética de los ocho años de vida del Círculo Argentino de Inventores,* a section on attendance at national and international conferences mentions those of the Liga Patriótica, which Ruiz Moreno attended with the Bachofen sisters in 1928 and 1929.

42. See *ICARM* 1:1–9 (Jan.–Sept. 1923), where this choice is explained as follows: "The official publication of the Círculo Argentino de Inventores will bear the first letters of the titles *Ingeniero* and *Coronel,* together with the first letters of the founder's name, as a perpetual homage begun during his lifetime, to be continued as a memorial, and as an inspiration to those who will be, in time, the successors of those who are making today's advances."

43. Between 1922 and 1928, the Círculo sponsored four contests for Argentine inventions, with the following categories: electricity and steam; construction; chemistry (a category declared to be without entries in the first contest and eliminated from those that followed); geodesics and topography (the field of Ruiz Moreno, who won first prize in the second contest); medicine; mechanics; agriculture (Elisa Bachofen's specialty); navigation and transportation; radio communication; and automotives. The third contest was devoted entirely to the construction of inexpensive roadways in coastal and similar regions (a topic that, to judge from the newspapers and technology publications, never appeared to hold any interest for amateur inventors).

44. Message from the Círculo, published on the occasion of its fifth anniversary; see *ICARM* 5:10–12 (Oct.–Dec. 1927).

Chapter 5

1. [An affluent suburb of Buenos Aires.—Trans.]

2. [Luna Park is an arena at the corner of Avenida Corrientes and Avenida Leandro N. Alem, in the San Nicolás district of Buenos Aires.—Trans.]

3. The photographs described here appeared on the covers of *Radio Cultura* 4–10 and 14.

4. *R.C.A.* 1 (Sept. 1924). A third-anniversary editorial in *R.C.A.* 3 summarized as follows the stages of the organization's brief history: "Three stages mark the development of our organization. After an era that the precursors of the organization rightly labeled heroic, the club was brought into being by a few amateurs, at a time when radio was largely unknown to the public; the club's own founders did not begin to suspect the interest that this marvelous invention would so quickly excite. It can be said that, among ourselves, radio was born along with the club, since a large contingent of radio amateurs got their start under the club's auspices. The organization made rapid progress, and its ranks swelled with the many members who wanted to learn the secrets of what was, for the majority, a mystery, especially with respect to what was involved in building radio sets, which were being sold at such high prices. . . . The second [stage] developed from the spread of circuits and lower costs for materials, but also from the boredom that these developments caused in those who were least persevering, and in those who had been seduced, at the very beginning, by a fad. At that point, demoralization took hold within our ranks, and although it did not attack the foundations of the club, it did result in a movement of protest and discontent; the steering committee was accused of negligence in the discharge of its duties. . . . This state of affairs did not last long; since its founding, the organization had become more and more firmly rooted, and the contingent of true radio amateurs was large enough for it to emerge victorious from this challenging time." But the crisis was not over, and significantly different projects for the organization were discussed in *R.C.A.* 3 and subsequent issues.

5. *Radio Cultura* 33 (Sept. 1923): "Remember, the *Radio Cultura* offices, at 600 Callao, Buenos Aires, have the world's best radios on permanent display, in addition to a salesroom for all your radiotelephony needs."

6. [For biographical information about the composer and performer Francisco Pracánico (1898–1971), along with sheet music and sound clips, see http://www.todotango.com/English/creadores/fpracanico.asp—Trans.]

7. The following fairly typical advertisement often appeared in issues of *Radio Revista*: "Exceptional offer this month—A complete receiver set, composed of 1 ebonite panel, perforated as you wish; 1 variocoupling; 1 variable capacitor; 1 set of pickup-coil spindles; 60 meters of .5 wire; 1 filament rheostat; 1 static resistor and capacitor; 1 telephone capacitor; 4 vacuum-tube sockets; 6 bronze connection terminals; 2 telephone connection terminals; 1 small roll of connective wire; 6 screws, with nuts; 12 terminals; 1 selector switch; small 2-meter rubber insulation tube. *All for*

35 pesos. Use these items to build a good receiver with capacity and range. Guaranteed to work, or your money back." *Crítica* published an article with instructions for assembling, at a cost of under 10 pesos, a receiver using a tube: "For amateurs with galena receivers, or for anyone who wants a radio, we offer this simple, economical receiver offering selectivity, range, and clarity"; see *Crítica*, Sept. 3, 1927. This receiver's success earned it the name Radio Crítica, and articles on its improvement continued to be published. In 1923, the paper had already published its "Radiodictionary," replete with technical instructions, and after August 1, 1927, it began publishing the "Weekly Radio Page," which alternated with the "Little Course on Radio." *Crítica* never abandoned its remarkably strong coverage of news about advances in Argentina and abroad. [See also chap. 3, n. 3.—Trans.]

8. *Radio Revista* 15 (Sept. 1923).

9. *Radio Revista* defended the inclusion of advertisements in its broadcasts as the only way of providing free service to its users and financing its program of broadcasts (a claim it had made as early as the first issue).

10. *Crítica*, Sept. 12, 1923.

11. *Crítica*, "Voice of the Air" section, Sept. 11, 1923.

12. *Crítica*, "Voice of the Air" section, Sept. 12, 1923.

13. *Crítica*, "Voice of the Air" section, Sept. 16, 1923. Several days later, an ingenious fellow made his appearance in *Crítica*'s "Voice of the Air" section, with an idea that was more practical than decorative: he had obtained superior results by substituting another metal for the block of galena; see ibid., Sept. 20, 1923. The same character wrote to several magazines, spreading the news about his alleged discovery but not specifying the metal with which the galena had been replaced; these magazines, more skeptical than *Crítica*, noted the omission.

14. Pablo Iturralde, "Lo que hace falta," *Radio Revista* 78 (Aug. 1926).

15. See, for example, *Radio Revista* 15.

16. There were at least four widely read manuals: those by Jorge A. Duclout, by Duclout and Guillermo Guntsche, by F. R. Pedraza, and by E. Repetto.

17. See, among dozens of examples (since one contest followed practically on the heels of another), the results published in *Radio Revista* 20 (Feb. 1924). Ten prizes were awarded—three for work on the use of a lighting current in a radio (which would eliminate the need for the huge batteries that the equipment had required until then), one for building a battery, one for quick solutions to emergencies, and the rest for building circuits, receivers, and microphones.

18. Editorial, *Radio Revista* 21 (Mar. 1924).

19. For example, the Radio Club was criticized for its inaction on regulations that governed the official recognition of wavelengths for radio amateurs. Until well into the 1920s, one of the official ordinances called for applicants to pass a test on Morse code, which was clearly irrelevant to radio broadcasting and only demonstrated the administrative backwardness that was justly criticized by the amateurs, who on this point, as well as in other matters that involved standing up to the authorities, felt very inadequately represented by the Radio Club Argentino. See *Radio Revista* 78 (Aug. 1926).

20. The expositions were held one after another, apparently with great success. A September 1923 expo on radiotelephony was followed in October by one on radiocommunications, which featured not only commercial booths but also a survey of early history that was sponsored by the Postal Service, and that exhibited the first telephone ever installed in Argentina, and the first Morse telegraph. Also for the October expo, an amateur historian had prepared a "panorama," very much in keeping with the aesthetic of the era, "that achieved, against the background of a rustic landscape, highly faithful reproductions of the aborigines' means of communication," by contrast with the Unión Telefónica's exhibit of all the types of telephones employed in Argentina up to that point, which the public could now use because they were now connected to the general network; see *Radio Revista* 16 (1923). The expos were held every year and received good coverage from both the specialized and the lay press. The army, together with the Unión Telefónica and commercial firms, participated in the exposition of 1925, using imported equipment on which local modifications had been successfully performed, as the exposition catalog pointed out; the Postal Service presented "a scale-model railroad, with telegraph wires running alongside it for sending messages." *Radio Revista* (Nov. 1925) criticized the admission price of 1 peso, which the magazine judged too steep for many of the amateurs. The exposition catalogs, which were booklets of some thirty to forty pages, mentioned all the commercial radio stores and their offerings as well as the Radio Club Argentino and the services it made available to its members; they also listed the wavelengths of several dozen ham radio operators and the radios that had been submitted for formal judging. By the end of the 1920s, the expos were already becoming an institution. One newspaper cited their success as if speaking of a familiar annual event, emphasizing that at the previous year's exposition the "huge crowds" had "surpassed many people's optimistic predictions"; see *El Mundo*, Apr. 30, 1930. In fact, again according to *El Mundo*, the exposition of 1929 had included not only radio but also cinematography, and

the exhibitors provided information to suggest that there had been eighty thousand visitors; see *El Mundo*, Dec. 12, 1929.

21. In the membership rolls of the Radio Club Argentino, the names and addresses of amateurs from all the working-class districts of Buenos Aires, and from a great many other regions around the country, appeared along with (and, after a few years, began to outnumber) surnames and addresses that indicated middle- and upper-middle-class status. For these lists of new members, see issues of *R.C.A.* published after 1924.

22. *R.C.A.* 11 (July 1925).

23. Among the new ones was *Ciencia Popular*, which appeared in August 1928 as part of a small empire of technical publications, and whose original launchpad had been *Radio Revista*.

24. See *El Mundo*, Nov. 19, 1928: "The number of radio stations has grown at an alarming rate, and today there are more than fourteen firms using the airwaves commercially, along with several others that broadcast from time to time. These radio stations have taken root in our city, broadcasting from dawn to midnight." The commentary went on to describe what were still primitive conditions in commercial radio broadcasting: "At present, the existing studios are provisional and improvisational, and they cannot be depended on to be perfectly soundproof. This is why honking sounds can be clearly heard on the afternoon broadcasts, when traffic is heavy and automobiles are making frequent use of their horns." The catalog for the May 11, 1930, exposition on radio, cinematography, and phonography announced the formation of a cooperative association for commercial broadcasters, self-defined as "cultural entities," which defended their role—made possible, they stated, by commercial advertising—in terms of community service. The president of the association was Antonio C. Devoto, and there were eighteen stations listed as having been established in the federal capital or surrounding areas.

25. *R.C.A.* 1 (1924): "Among the many opinions offered by our members, we note a certain level of paired opposition. Some believe the club should acquire a powerful transmitter of its own, to broadcast concerts and the like; others think the club's mission should be to take on the transmitters that already exist, and to serve as a kind of airwave police overseeing amateur radio broadcasts. Some want to see the formation of a cooperative, to ensure that the materials and equipment for building radios can be obtained at a reasonable price; others think the club's mission is to provide a radio set to every member who doesn't own one. And there is no shortage of those who want to see a large number of radios installed in our meeting space, for listening to the daily concerts." There was chaos on wavelengths that were already occupied, had just been assigned, or were

being usurped, and the problem not only persisted but got worse as time went on; see the open letter to the director of the telegraph agency that appeared in *R.C.A.* 44 (Mar. 1929).

26. By 1930, they were already heavily represented in *El Mundo*'s Buenos Aires movie listings.

27. *Ciencia Popular* 22 (May 1930).

28. In this connection, see Rafael Filippelli, "Imágenes de *Lulu*," *Lulú, revista de teoría y técnicas musicales* 1 (Aug. 1991).

29. *Crítica*, Feb. 13, 1924.

30. [Sofía Bozán (1904–1958), born in Buenos Aires, was a popular actress in musical theater and was also a recording star.—Trans.]

31. *Ciencia Popular* 3 (Oct. 1928).

32. There was an abundance of such information in the daily press and in almost every issue of *Ciencia Popular* from 1929 to 1930.

33. Advertised in *Crítica*, Oct. 28, 1930.

34. *El Mundo*, Nov. 6, 1929.

35. The difference is revealed by the cost of a used movie camera alone, even without the expense of film and developer. A used Kodak camera might run as high as 300 pesos, the price of a ready-made deluxe radio; a hand-cranked Pathé-Baby camera, also used, cost 50 pesos; another model, this one with a windup motor, cost 190 pesos. Then, of course, there was the need for a projector, a screen, lighting equipment, film, developer, and so on, none of which could be homemade. The fact that the same items were offered for sale month after month in the classified ads of *Ciencia Popular* suggests that the market for them was totally flat. And, apart from the price of the camera and the film (not to mention the projector), there was the added difficulty that none of this equipment, unlike the equipment used to build a radio, could be obtained or repaired on a part-by-part basis. Cinematic technology was not for amateurs.

36. See *Ciencia Popular* 7 (Feb. 1929). Israel Chas de Cruz was a chronicler of the movies for many decades, and Alberto Etchebehere served as lighting designer for dozens of Argentine films until the 1960s.

37. *Crítica*, Mar. 5, 1928.

38. *Crítica*, Dec. 16, 1926.

39. *Crítica*, Feb. 9, 1928.

40. *Crítica*, Apr. 9, 1928: "Engineers who have examined this new device estimate that it can be built for less than $400, and they also say that a skilled amateur should be able to build it himself, since all the necessary items except a special tube can be found these days in radio stores and other shops."

41. See "La televisión al día," *El Mundo*, Jan. 22, 1930. This article,

which was fairly precise in defining the principal technical problem to be solved (synchronization between the images emitted and the images received), was illustrated with a photograph that was 2 centimeters high by 2 centimeters wide, the original size of the image produced by a television set at General Electric's North American laboratories.

42. *Ciencia Popular* 1 (Aug. 1928).

43. See ibid. for a passing reference: "In patent registration offices all over the world, there are countless plans for television sets, but not one of them can be considered definitive. This is why we say that every talented person can make a contribution to solving the problem."

44. *Ciencia Popular* 15 (Oct. 1929). *La Nación* had already dealt with the device that was built by Duclout and exhibited at the Teatro Opera exposition. The paper noted the rudimentary, experimental nature of the broadcast as well as the achievement it represented; see *La Nación*, May 19, 1929. The corresponding coverage in *Crítica* was somewhat more sensational. At any rate, both papers published photographs of the device and diagrams of how it worked, and these were the same ones that appeared in subsequent issues of *Ciencia Popular.*

45. A number of articles were devoted to the telegraphic broadcasting of images—see, for example, *Ciencia Popular* 4 (Nov. 1928)—and the newspapers of the day dealt with this advance primarily in terms of its journalistic applications.

Chapter 6

1. On this topic, and particularly on the popularization of Freud, see Hugo Vezzetti, ed., *Freud en Buenos Aires, 1910–1939* (Buenos Aires: Puntosur, 1989).

2. Dr. E. E. Fournier D'Albe, "La trasmisión del pensamiento por T.S.H.," *Ciencia Popular* 1 (Aug. 1928).

3. See Hugo Vezzetti, ed., *El nacimiento de la psicología en la Argentina* (Buenos Aires: Puntosur, 1988).

4. See *Ciencia Popular* 3 (Oct. 1928), and *Ciencia Popular* 7 (Feb. 1929). The course was described as "experimental" and included sections on hypnotism and suggestion.

5. Conan Doyle also wrote a number of science fiction stories, including "The American Tale" (1879), "The Adventure of the Creeping Man" (1903), "The Great Brown Pericord Motor" (1911), and "The Horror of the Heights" (1913), as well as stories on reincarnation and parapsychological communication under hypnosis, such as "The Great Keinplatz Experiment" (1894). See Charles Waugh and Martin Greenberg, eds., *The*

Best Science Fiction of Arthur Conan Doyle (Carbondale: Southern Illinois University Press, 1981).

6. *Crítica*, May 21, 1922.

7. *Crítica*, Mar. 7, 1925: "There are serious, educated individuals who have dedicated themselves, with all their ability and enthusiasm, to delving into this mysterious science."

8. [See chapter 2, n. 30.—Trans.]

9. See *Crítica*, Sept. 28, 1927.

10. See *Crítica*, Mar. 7, 1925: "Spiritism has become quite important in Argentina." See also the long article in *Crítica* magazine, Nov. 22, 1926. Articles of this kind appeared at least once a month throughout the decade. Theosophy was another topic frequently covered in the paper, and in April 1928 *Crítica* published an article several pages long that explained its general principles, starting from its origins in ancient Greece and taking the reader up to the contemporary schools. In July 1930, another very long article expounded the ideas of Helena Blavatsky and her local followers (who were objects of Roberto Arlt's fury and fascination in "Las ciencias ocultas en la ciudad de Buenos Aires").

11. *Crítica*, Sept. 28, 1927.

12. See "La ciudad está minada de curanderas, curanderos y adivinas que explotan la credulidad del pueblo," *Crítica*, Jan. 21, 1929. See also Roberto Arlt, "El hombre de las ciencias ocultas," *El Mundo*, Jan. 3, 1930, which depicts one of the strange but typical characters who were normally seen around Buenos Aires, often in the company of some failed inventor.

13. "El profesor Bernardo, con el medium Alfredi de fama mundial, harán sus grandes experiencias científicas de esoterismo e hipnotismo," *Crítica*, July 26, 1930. [The French physiologist Charles Robert Richet (1850–1935), winner of the Nobel Prize in 1913, was deeply interested in hypnosis and extrasensory perception. The French neurologist Jean-Martin Charcot (1825–1893) was renowned for his work on hypnosis and hysteria.—Trans.]

14. The head died at eight o'clock; see *Ciencia Popular* 8 (Mar. 1929). See also *Ciencia Popular* 14 (Sept. 1929), which reprinted an article from *El Día Médico* in which a Russian scholar explained how an artificial pump was used to establish blood circulation outside the body. Devices for resuscitating the dead were the topic of several articles in the 1920s; see, for example, the one in *Crítica*, Sept. 12, 1929: "The invention is based on the principle of stimulating the heart muscles by inserting an electrified needle into the heart and holding it there for ten seconds. The inventor cited the case of a little girl who was resuscitated after having been pronounced dead. It is hoped that this new device will be of great help to people who have drowned."

15. Horacio Quiroga also dealt with such themes in the first years of the twentieth century; see chapter 1, "Horacio Quiroga and Techno-scientific Theory." Literary critics in the area of science fiction have often noted the Faustian impulse that informs these experiments, just as they have pointed to the figure of the scientist who breaks natural and moral laws. See L. Sprague de Camp and Thomas Clareson, "The Scientist," in P. Warrick, M. Greenberg, and H. Olander, *Science Fiction: Contemporary Mythology* (New York: Harper and Row, 1978), and Robert Philmus, *Into the Unknown: The Evolution of Science Fiction from Godwin to Wells* (Berkeley: University of California Press, 1970), 82ff.

16. The article gave details; see *Crítica*, Oct. 19, 1927: "Nine female chimpanzee specimens were brought here from Africa by Professor Tobolkin, director of the Moscow Anthropological Institute. The chimps will be free to roam about the garden, which is surrounded by a high fence, where the climate, the food, and the atmosphere will made to resemble, as much as possible, the areas of the African jungle where they were born and grew to adulthood. . . . The chimps will be fertilized scientifically, with human sperm, and a group of scientists will monitor the development of the pregnancies." Other articles, equally fascinating, belong to the same science fictional environment; see *Crítica*, Mar. 30, 1924, for a report about an Austrian scientist who caused eyes to grow in the heads of eyeless subterranean rats. See also *Crítica*, Jan. 21, 1925, for assurances that the day was coming when science would discover how married couples could determine the sex of their offspring.

17. See *Crítica*, Jan. 31, 1927: "After twenty years of intense and fruitless research, life has been created with the use of a strange substance that produces spontaneous generation in the seeds of life that remain in a latent state within the egg. This substance, extracted from the blood of various animals, has the power to fertilize new individuals from the egg of a sea urchin. . . . According to Dr. Clark's report, 'These chemically fertilized animals do not live longer than two or three days. It seems that we need to provide some element that is not supplied by our physiomechanical method of fertilization so that we can obtain adequate fertilization after the membrane has begun to form and the developmental process has begun.'"

18. *Crítica*, Apr. 8, 1928.

19. *Crítica*, Mar. 2, 1923.

20. See, for example, *El Mundo*, Dec. 21, 1928; Roberto Arlt, "El gremio de las curanderas y santeras," *El Mundo*, July 29, 1928; and the comprehensive report published in *Crítica*, Jan. 21, 1929.

21. Before its use in Argentina, the Voronoff method was used in a French experimental surgery laboratory (even though its inventor was

Russian), and from there it spread all over the world, thanks to the wire services' appetite for news of this kind.

22. The whole affair was covered by *Crítica* in April, May, and August of 1926 and in March, April, and May of 1928 as well as by *El Mundo* in February of 1929.

23. *El Mundo*, May 15, 1928.

24. *El Mundo*, Apr. 24, 1930. [Isidore Marie Auguste François Xavier Comte (1798–1857) developed the philosophy known as positivism and is remembered for coining the term "sociology" and applying the scientific method to the social world.—Trans.]

25. "Asuero y la medicina," *El Mundo*, June 16, 1930.

26. There is a strong resemblance between Asuero's social whirl and that of the ill-fated Professor Landormy in Arturo Cancela's *Historia funambulesca del profesor Landormy*, a story published in Buenos Aires in 1944 but set in the mid-1920s.

27. See, for example, the interview with Dr. Eduardo Casterán in *El Mundo*, June 21, 1929, at a time when the local press as a whole, *El Mundo* included, was publishing news about the unusual cures that Asuero had brought about in Europe.

28. *El Mundo*, Apr. 30, 1930. On Asuero's technique and its predecessors, see also *Ciencia Popular* 12 (July 1929).

29. [Chapadmalal is twelve miles south of Mar del Plata.—Trans.]

30. See *Crítica*, May 30, May 31, June 1, June 2, June 4, June 6, June 10 (the front page is devoted to the subpoena), July 11, and July 14, 1930.

31. The survey began on April 19, 1926, with a response from Bernardo Houssay, followed between April 20 and May 7 by responses from Héctor Dasso, Angel Roffo, Antonio Agudo Avila, Joaquín Llambías, Julio Iribarne, Francisco Rophille, Abel Zubizarreta, Nerio Rojas, Avelino Gutiérrez, Pedro Escudero, Gerónimo Bosch Arana, and Juan José Vitón. *Crítica* conducted another survey in 1929, this one focused on a particular question: How far could science go in experimenting on patients? Roffo, Llambías, and Rojas were interviewed again for this survey, as was an equally distinguished group that included Carlos Cometto, Guillermo Bosco, Alfredo Lanari, Elbio Fernández, Salvador Mazza, and Osvaldo Loudet, among others. [With Carl and Gerty Cori, the physiologist Bernardo Houssay went on to win the Nobel Prize in 1947.—Trans.]

32. See, for example, the following articles, in which Asuero's method was denounced as "a product of the abnormal situation in Spain during the dictatorial regime" and as "a simple hypnotic procedure," and where the doctor himself was condemned while the toll of his relapsing patients mounted: "Histerismo, anormalismo, curandería, dicen de la asueroterapia tres maestros españoles. Unamuno, Marañón y Tapia convienen en que es

un producto de la situación anormal de España durante el régimen dictatorial," *Crítica*, May 20, 1930; "Avelino Gutiérrez, ilustre médico español, condena enérgicamente el caso Asuero," *Crítica*, May 23, 1930; "Maestros médicos argentinos hablan del método Asuero. No quieren molestarse en combatir lo que morirá automáticamente. Es un sencillo procedimiento de sugestión," *Crítica*, May 28, 1930; and "Son numerosos los enfermos 'curados' por el método Asuero que se han reagravado," *Crítica*, May 29, 1930.

The authorized representative in the EU for product safety and compliance is:
Mare Nostrum Group
B.V Doelen 72
4831 GR Breda
The Netherlands

www.ingramcontent.com/pod-product-compliance
Lightning Source LLC
LaVergne TN
LVHW091514200926
848483LV00005B/416/J
9780804735421